Springer Theses

Recognizing Outstanding Ph.D. Research

Aims and Scope

The series "Springer Theses" brings together a selection of the very best Ph.D. theses from around the world and across the physical sciences. Nominated and endorsed by two recognized specialists, each published volume has been selected for its scientific excellence and the high impact of its contents for the pertinent field of research. For greater accessibility to non-specialists, the published versions include an extended introduction, as well as a foreword by the student's supervisor explaining the special relevance of the work for the field. As a whole, the series will provide a valuable resource both for newcomers to the research fields described, and for other scientists seeking detailed background information on special questions. Finally, it provides an accredited documentation of the valuable contributions made by today's younger generation of scientists.

Theses are accepted into the series by invited nomination only and must fulfill all of the following criteria

- They must be written in good English.
- The topic should fall within the confines of Chemistry, Physics, Earth Sciences, Engineering and related interdisciplinary fields such as Materials, Nanoscience, Chemical Engineering, Complex Systems and Biophysics.
- The work reported in the thesis must represent a significant scientific advance.
- If the thesis includes previously published material, permission to reproduce this must be gained from the respective copyright holder.
- They must have been examined and passed during the 12 months prior to nomination.
- Each thesis should include a foreword by the supervisor outlining the significance of its content.
- The theses should have a clearly defined structure including an introduction accessible to scientists not expert in that particular field.

More information about this series at http://www.springer.com/series/8790

Yusuke Kawamoto

Synthesis and Biological Evaluation of Pyrrole–Imidazole Polyamide Probes for Visualization of Telomeres

Doctoral Thesis accepted by
Kyoto University, Kyoto, Japan

Springer

Author
Dr. Yusuke Kawamoto
Department of Chemistry and Biochemistry
University of California
San Diego, La Jolla, CA, USA

Supervisor
Prof. Hiroshi Sugiyama
Department of Chemistry
Graduate School of Science
Kyoto University
Kyoto, Japan

ISSN 2190-5053 ISSN 2190-5061 (electronic)
Springer Theses
ISBN 978-981-13-6911-7 ISBN 978-981-13-6912-4 (eBook)
https://doi.org/10.1007/978-981-13-6912-4

Library of Congress Control Number: 2019932715

This Springer imprint is published by the registered company Springer Nature Singapore Pte Ltd.
The registered company address is: 152 Beach Road, #21-01/04 Gateway East, Singapore 189721, Singapore

Supervisor's Foreword

Over the several decades, varied molecules to target specific sequences on nucleic acids have been developed. They have been used for therapeutics of cancer and other diseases, visualization of specific regions, regulation of genes, and so on. In this area, we have worked on synthetic molecules recognizing DNA-specific sequences, named pyrrole (Py)-imidazole (Im) polyamides.

In this thesis, Dr. Yusuke Kawamoto reports Py-Im polyamides targeting human telomeres. The author demonstrates polyamides targeting longer base pairs than the previously reported ones and their cellular application. In Chap. 1, the author discusses the history and recent studies of Py-Im polyamides to help readers to learn the background. In Chap. 2, the author reports the development of the facile synthetic method of tandem hairpin Py-Im polyamides targeting 12 bp in human telomere repetitive sequence and the method to visualize telomeres under mild condition without the harsh denaturation. In Chap. 3, the author reports longer tandem trimer Py-Im polyamides targeting 18 bp in telomere sequence and shows that extending the target base pairs decreases the background signals of fluorescently labeled polyamides and increases the specificity. In Chap. 4, tandem tetramer Py-Im polyamides targeting 24 bp are reported. 24 bp is the longest reported binding site for synthetic, non-nucleic-acid-based, sequence-specific DNA-binding molecules. Visualization of telomeres in cells and high-throughput sequencing technique demonstrate that tandem tetramer polyamides have higher specificity to telomeres than shorter polyamides as discussed in Chaps. 2 and 3.

This thesis clearly demonstrates the development and optimization of tandem hairpin Py-Im polyamides and visualization of telomeres in cells using the polyamides. I believe this can be applied to future studies on DNA-binding molecules and their cellular applications.

Kyoto, Japan
November 2018

Prof. Hiroshi Sugiyama

Parts of this thesis have been published in the following journal articles:

1. "Sequence-specific DNA Binding Pyrrole–Imidazole Polyamides and Their Applications" Kawamoto, Y.; Bando, T.; Sugiyama, H. *Bioorg. Med. Chem.* **2018**, *26*, 1393–1411.
2. "Development of a New Method for Synthesis of Tandem Hairpin Pyrrole-Imidazole Polyamide Probes Targeting Human Telomeres" Kawamoto, Y.; Bando, T.; Kamada, F.; Li, Y.; Hashiya, K.; Maeshima, K.; Sugiyama, H. *J. Am. Chem. Soc.* **2013**, *135*, 16468–16477.
3. "Tandem Trimer Pyrrole-Imidazole Polyamide Probes Targeting 18 Base Pairs in Human Telomere Sequences" Kawamoto, Y.; Sasaki, A.; Hashiya, K.; Ide, S.; Bando, T.; Maeshima, K.; Sugiyama, H. *Chem. Sci.* **2015**, *6*, 2307–2312.
4. "Targeting 24 bp within Telomere Repeat Sequences with Tandem Tetramer Pyrrole–Imidazole Polyamide Probes" Kawamoto, Y.; Sasaki, A.; Chandran, A.; Hashiya, K.; Ide, S.; Bando, T.; Maeshima, K.; Sugiyama, H. *J. Am. Chem. Soc.* **2016**, *138*, 14100–14107.

Acknowledgements

First, I would like to deeply appreciate my research supervisor, Professor Hiroshi Sugiyama, for teaching me about science and valuable suggestions about my research. His guidance and discussions helped me to solve various problems. I also deeply appreciate Associate Professor Toshikazu Bando for teaching me about chemical synthesis and his carefully considered feedback and valuable comments and suggestions, especially about the design of compounds and experimental plans. I would also like to appreciate other Sugiyama group staffs, Associate Professors Masayuki Endo and Asako Yamayoshi, and Assistant Professors Soyoung Park and Namasivayam Ganesh Pandian for their helpful advice and discussions.

I would like to thank senior laboratory members, Dr. Gengo Kashiwazaki, Dr. Yuki Suzuki, and Dr. Hironobu Morinaga, for their coaching of me.

I am grateful to Ms. Yasuko Niimi, Ms. Takako Futamata, Ms. Chikako Bamba, Ms. Kumi Hidaka, and Ms. Tomoko Emura for their kind support of my study as secretaries. I am also grateful to laboratory technicians, Ms. Kaori Hashiya for performing solid-phase synthesis, and Mr. Shinsuke Sato for helping me in cell culturing.

I would like to express my special thanks to laboratory mates, Dr. Anandhakumar Chandran for performing high-throughput sequencing and its analysis, Dr. Yue Li for supporting the synthesis of our compounds, Dr. Junichi Taniguchi for helpful discussions and suggestions about our compounds and their cellular evaluations, and Mr. Sefan Asamitsu for supporting SPR experiments and helpful discussions about G-quadruplexes. I am also grateful to all the members of the Sugiyama's laboratory who supported and gave comments about my work.

I am very much indebted to our collaborators Professor Kazuhiro Maeshima, Assistant Professor Satoru Ide, Ms. Asuka Sasaki, and Ms. Fukumi Kamada (National Institute of Genetics) for cellular experiments, capturing excellent cell images, and their helpful discussions. Thanks to their data, we successfully published many excellent papers, some of which are included in my thesis. I would like to thank Professor Peter B. Dervan (California Institute of Technology) for welcoming me to visit his laboratory and valuable discussion and suggestion about the design of pyrrole-imidazole polyamides. I also appreciate Assistant Professor Shigeo S. Sugano (Ritsumeikan University) for evaluating our compounds using plants and helpful suggestion about analysis of our high-throughput sequencing data.

I would like to thank Professor Hanbin Mao, Mr. Shankar Mandal, and Mr. Mohammed Enamul Hoque (Kent State University) for their teaching and helpful discussions about the research of G-quadruplex using their optical tweezers and our compounds and helping a lot all about my staying in the USA.

I also thank Japan Society for the Promotion of Science for financial support (Research Fellowship for Young Scientists), our Department of Chemistry for financial support of overseas collaborative research, and WPI-iCeMS for financial support of visiting laboratories in the USA.

Finally, I will give a deep appreciation to my family for their constant support and encouragement of my motivation.

Contents

Chapter 1
Introduction: Sequence-Specific DNA Binding Pyrrole–Imidazole Polyamides and Their Applications

Abstract Pyrrole–imidazole polyamides (Py–Im polyamides) are cell-permeable compounds that bind to the minor groove of double-stranded DNA in a sequence-specific manner without causing denaturation of the DNA. These compounds can be used to control gene expression and to stain specific sequences in cells. Here, we review the history, structural variations, and functional investigations of Py–Im polyamides.

Keywords DNA binding molecules · Pyrrole–imidazole polyamides · Minor groove binder · Gene regulation · Cellular imaging

1.1 Introduction

DNA is a well-known biomolecule that encodes genetic information in living cells and usually has a B-form right-handed double-helical structure [1], which has major and minor grooves. In addition, it can form noncanonical structures such as G-quadruplexes in living cells [2]. Targeting DNA with synthetic molecules has the potential to regulate gene expression and may be applicable to diagnosis and therapy of human diseases including cancer and infection. It also has potential for use in gene regulation research without the need for gene modification or construction and transfection of genetic material. Many researchers, including our group, have studied pyrrole (Py or P)–imidazole (Im or I) polyamides, abbreviated as Py–Im polyamides, PIPs or PAs, which bind to the minor groove of double-stranded canonical B-DNA in a sequence-specific manner.

Py–Im polyamides were developed based on the natural antibiotics netropsin and distamycin A [3]. These natural molecules contain two and three Py moieties, respectively (Fig. 1.1). Netropsin preferentially binds to the minor groove of A/T-rich

This chapter is reprinted and modified with permission from "Sequence-specific DNA Binding Pyrrole–Imidazole Polyamides and Their Applications" Kawamoto, Y.; Bando, T.; Sugiyama, H. *Bioorg. Med. Chem.* **2018**, *26*, 1393–1411. Copyright 2018 Elsevier.

Y. Kawamoto, *Synthesis and Biological Evaluation of Pyrrole–Imidazole Polyamide Probes for Visualization of Telomeres*, Springer Theses,
https://doi.org/10.1007/978-981-13-6912-4_1

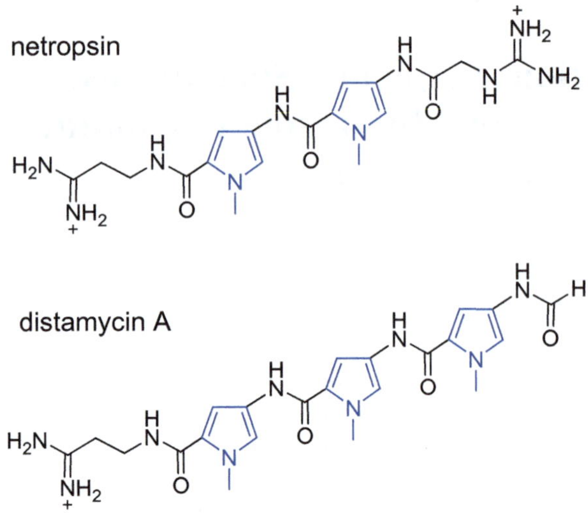

Fig. 1.1 Chemical structures of netropsin and distamycin A. The image is quoted from Ref. [10i]. Copyright 2018 Elsevier

double-stranded B-DNA with a 1:1 stoichiometry [4]. Distamycin A also binds to the minor groove of A/T-rich B-DNA as 1:1 or 2:1 ligand–DNA complexes [5]. These molecules interact with DNA via hydrogen bonds between the amide NH group and adenine N3 or thymine O_2, and via van der Waals interactions between C3–H on the Py moiety and the C2–H of adenine. Based on structural analysis, Dickerson and coworkers hypothesized that substitution of N for C3–H in the *N*-methylpyrrole moiety, i.e., replacing *N*-methylpyrrole with *N*-methylimidazole, would generate a hydrogen bond between the exocyclic amino group on guanine and N3 on imidazole and permit recognition of G/C base pairs [4]. This model, based on a 1:1 stoichiometry, would not distinguish GC from CG, but rather G/C from A/T.

In 1992, Dervan and coworkers demonstrated that a netropsin analogue ImPyPy-dimethylaminopripylamide (Dp) recognized and bound to 5′-TGTCA-3′ as a 2:1 ligand–DNA complex (Fig. 1.2) [6]. The polyamide formed an antiparallel homodimer in the minor groove of DNA and bound with the N- to C-terminus aligned in the 5′–3′ direction. Importantly, the study demonstrated the fundamental rules of DNA recognition by precisely oriented pairs of pyrrole and imidazole connected by amide bonds: the Im/Py pair recognizes G/C base pairs, Py/Im recognizes C/G, and Py/Py targets A/T and T/A (Fig. 1.2). ImPyPyDp could also form a heterodimer with distamycin A and recognize 5′-TGTTA-3′ [7]. Subsequent studies by various groups have extended the applicability of Py–Im polyamides. In this review, we focus on the development and application of Py–Im polyamides.

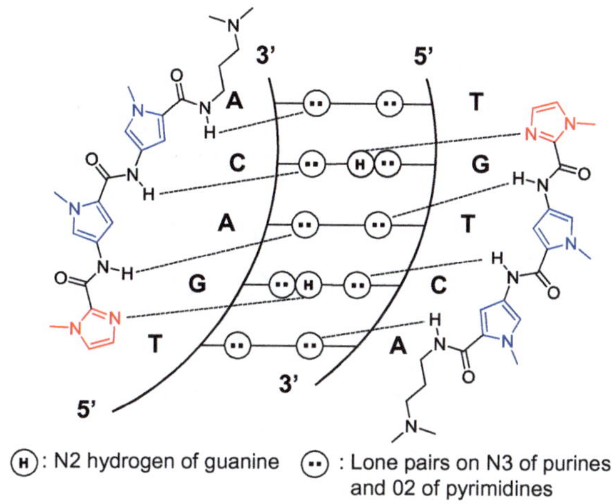

Fig. 1.2 Molecular recognition of the minor groove of dsDNA by a netropsin derivative ImPyPyDp [6]. The image is quoted from Ref. [10i]. Copyright 2018 Elsevier

1.2 Hairpin Structure and Pairing Rule of Py–Im Polyamides

The short linear Py–Im polyamides described in the previous section can form homo- and heterodimers in the minor groove of dsDNA, and recognize specific sequences [6, 7]. Based on this finding, covalent linking dimers named H-pin polyamides have been developed, in which two linear oligomers are connected via the nitrogen of the central pyrrole ring by an alkyl linker. H-pin polyamides have stronger affinity and greater specificity for their matching sequences compared with the unlinked parent linear monomer polyamides [8].

To increase the variety of covalent Py–Im dimeric polyamides, Dervan and coworkers demonstrated a hairpin motif, where the C-terminus of one polyamide and the N-terminus of another moiety are covalently connected by γ-aminobutyric acid (γ-turn) (Fig. 1.3). This linking enhanced their binding affinity and specificity while maintaining their binding orientation (N to C with respect to the 5′–3′ direction of the DNA) [9]. Straightforward solid-phase synthetic methodology using Boc [9b] and Fmoc chemistry [9c] improved the efficiency of synthesis of the hairpin Py–Im polyamides and conferred programmability. This method dramatically extended the range of Py–Im polyamides and accelerated their investigation. Furthermore, the binding affinity and specificity of eight-ring hairpin Py–Im polyamides for a 6-bp target sequence were found to be comparable to those of DNA binding proteins (K_D ~ 10^{-9} nM) [9d]. Currently, hairpin motifs are the most commonly used in studies of Py–Im polyamides [10].

Structural studies confirmed the fundamental pairing rules for Py–Im polyamides: i.e., Im/Py recognizes G/C and Py/Py recognizes A/T or T/A through hydrogen bonds between the N2-H of guanine and the N3 of Im, the N3 of guanine and the amide-NH of Im, the O_2 of cytosine and the amide-NH of Py, the N3 of adenine and the

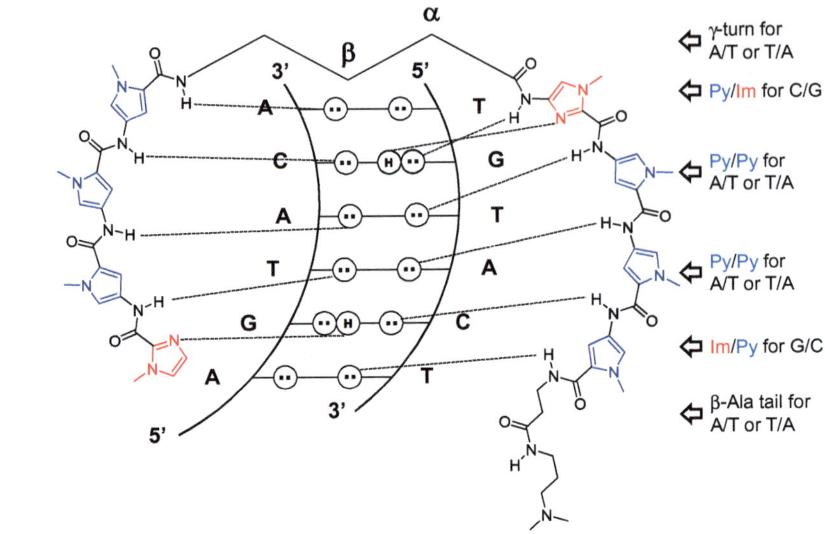

Fig. 1.3 Molecular recognition of dsDNA by a hairpin Py–Im polyamide. The image is quoted from Ref. [10i]. Copyright 2018 Elsevier

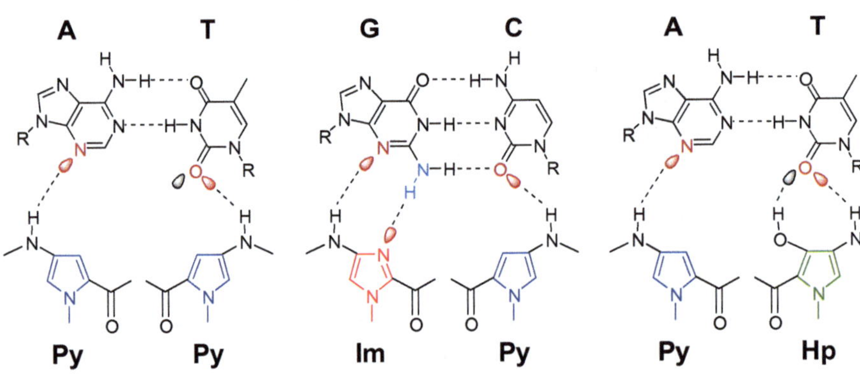

Fig. 1.4 Pairing rules for DNA-binding polyamides indicating the hydrogen bonds between bases and Py, Im or Hp residues. The image is quoted from Ref. [10i]. Copyright 2018 Elsevier

amide-NH of Py, and the O_2 of thymine and the amide-NH of Py (Figs. 1.3 and 1.4) [5b, 11, 12]. In addition, the Im/Im pair can target a T/G mismatch [13].

Subsequent studies showed that the γ-turn works as a recognition unit for A/T or T/A (Fig. 1.3) [14]. Modification of the α-position of the γ-turn by an (*R*)-amino or acetyl amino group enhanced its binding affinity and disfavored the reverse binding mode (C to N direction of Py–Im polyamides with respect to the 5′–3′ direction of DNA) while maintaining A/T or T/A recognition [15, 16]. In contrast, introduction of an (*S*)-amino group to the α-position of the γ-turn decreased its binding affinity to natural D-DNA compared with that of unsubstituted polyamides, but a single-molecule force spectroscopy analysis demonstrated that the (*S*)-amino group induced preferential binding to L-DNA, i.e., the mirror-image of natural D-DNA [17]. Modification of the β-position of the γ-turn also enhanced the binding affinity, but this modification favored reverse binding [16, 18].

β-alanine inserted at the C-terminal enhanced both the binding affinity and specificity of the compound and recognized A/T or T/A (Fig. 1.3) [14, 19]. Although the upper limit of contiguous rings in Py–Im polyamides is five and more than five contiguous rings reduced the binding affinity and specificity because of overcurvature, β-alanine inserted within hairpin moieties works as a "spring" that relaxes the curvature of the polyamides to allow their efficient binding; this in turn allowed the extension of the range of target bases. β-alanine can form β/β, Py/β, and Im/β pairs as a substitute for Py rings. β/β and Py/β recognize A/T or T/A and Im/β recognizes G/C [20].

Im/Py can distinguish G/C from C/G, but the Py/Py pair cannot distinguish T/A and A/T. Dervan and coworkers reported that substitution of C3–H by C3–OH within a Py/Py pair generated an *N*-methyl-3-hydroxypyrrole (Hp)/Py pair. The Hp/Py pair could distinguish T/A from A/T (17- to 77-fold difference in affinity), C/G and G/C and, within the pair, Hp could bind to T through hydrogen bonds between the inserted hydroxyl group and the O_2 of thymine and the amide-NH and O_2 of thymine (Fig. 1.4) [21]. DNA-binding polyamides including Im/Py, Py/Im, Hp/Py and Py/Im pairs could distinguish all four Watson–Crick base pairs, but the use of Hp was limited because of its instability in both solid form and solution [10b, c, g].

Many studies have aimed to expand the repertoire of pairing and distinguish T/A from A/T by procedures such as modification of aliphatic linkers and design of alternative heterocycles. Dervan's group reported that α-substituted-β-amino acid (*S*)-isoserine (^{S}Is) pairs with β in the hairpin motif and that the ^{S}Is/β pair preferred T/A over A/T with 7.2 times greater affinity (Fig. 1.5a) [22a]. Boger and coworkers demonstrated that (*R*)-α-methoxy-β-alanine ($\beta^{(R)-\mathrm{OMe}}$) as a substitute for the γ-turn maintained the binding affinity and preferentially adopted a hairpin versus a linear binding mode (Fig. 1.5b) [22]. Our group also studied the effect of modification of the aliphatic residues on recognition and found that (*S*)-α-hydroxyl-γ-aminobutyric acid ($^{S}\alpha_{\gamma\text{-OH}}$) and (*S*)-β-hydroxyl-γ-aminobutyric acid ($^{S}\beta_{\gamma\text{-OH}}$) in the hairpin motif could pair with β and that the substituted residues recognized T with 19- and 38-fold greater affinity, respectively, than A (Fig. 1.5c, d) [22c, d]. In addition, introduction of an (*S*)-hydroxyl group to the α-position of the γ-turn decreased the binding affinity but enabled this γ-turn to distinguish A/T from T/A with a 25-fold increase in affinity (Fig. 1.5e) [22e].

Recognition by aromatic five-membered heterocycles incorporated into the hairpin as substitutes for Py, Im and Hp (Fig. 1.6) has also been investigated [23]. 3-Pyrazolecarboxylic acid (3Pz) at the N-terminus paired with Py favored recognition of G/C similarly to Im/Py [23a]. In contrast, N-terminus 3-chlorothiophen (Ct) or 3-methoxythiophen (Mt) paired with Py could distinguish T/A from A/T with 3.5 and 6.3 times greater affinity, respectively [23b]. N-terminal 5-isopropyl thiazole (Nt) paired with Py preferentially bound G/C, similarly to Im/Py [23c]. This Nt compound could also recognize G within a short, linear polyamide called thiazotropin A

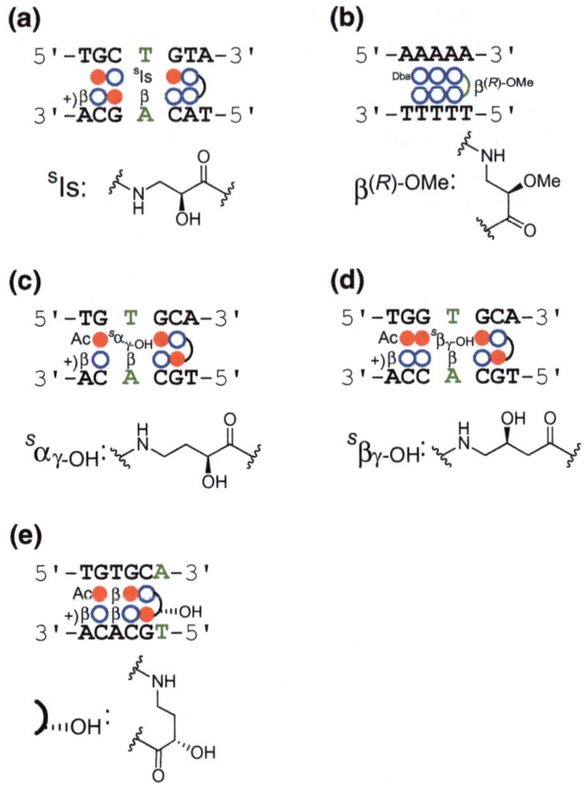

Fig. 1.5 Recognition of bases by modified aliphatic linkers. The image is quoted from Ref. [10i]. Copyright 2018 Elsevier

in a 2:1 ligand–DNA complex [23d, e]. *N*-methylpyrrazole (Pz)/Py and 1*H*-pyrrole (Nh)/Py pairs inside the hairpin motif bound A/T and T/A [23f, g]. Furan (Fn) incorporated into the hairpin paired with Py and Im could function as a Py mimic [23h]. 3-Methylthiophene (Tn) and thiophene (Tp) paired with Py within the hairpin preferentially bound T/A over A/T with a modest 3-fold greater affinity [23g].

Benzimidazole derivatives within hairpin polyamides have also been investigated as DNA recognition units [24]. Imidazopyridine (Ip), benzimidazole (Bi) and hydroxybenzimidazole (Hz) fused with the five-membered rings Py, Im, Ct or oxazole (No) without an intermediate amide bond could be introduced into the N-terminus and the inside of the hairpin polyamide backbone, and Ip, Bi and Hz could function as one recognition unit by pairing with Py or Im (Fig. 1.7). Ip/Py could distinguish G/C from C/G, A/T and T/A [24a]. Hz/Py could recognize T/A base pairs, mimicking Hp/Py [24a, c, d]. Bi behaved like Py; therefore, Bi/Py bound both A/T and T/A, and Bi/Im recognized C/G [24a, b]. Furthermore, Hz could pair with Bi to form a six-member/six-member ring-ring Hz/Bi pair, which could distinguish T/A from A/T and other base pairs [24c].

Structure	Recognition
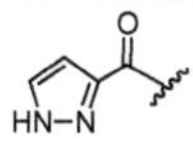 **3Pz**	3Pz/Py for G/C
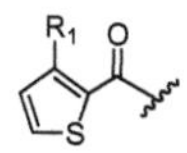 **Ct**: R_1 = Cl **Mt**: R_1 = OMe	Ct/Py and Mt/Py for T/A
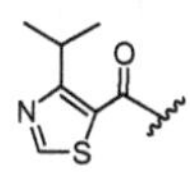 **Nt**	Nt/Py for G/C
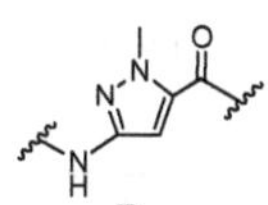 **Pz**	Pz/Py for A/T and T/A
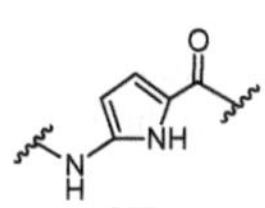 **Nh**	Nh/Py for A/T and T/A
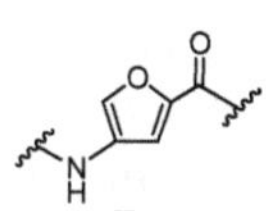 **Fn**	Fn/Py for A/T and T/A Fn/Im for C/G
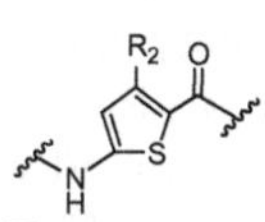 **Tn**: R_2 = Me **Tp**: R_2 = H	Tn/Py and Tp/Py for T/A

Fig. 1.6 Recognition of five-membered aromatic heterocycle pairs. The image is quoted from Ref. [10i]. Copyright 2018 Elsevier

Lee, Wilson and coworkers investigated the effect of substitutions at the N-terminus on the binding of linear triheterocyclic polyamides that were able to form homodimers in the minor groove of DNA [25]. The N-terminal formamido group present in distamycin is difficult to introduce into hairpin polyamides because it is difficult to synthesize [26]. Introduction of this formamido group was found to enhance the binding affinity and induce a change in the shift-stacking between two antiparallel monomers from fully overlapped to staggered, resulting in a change in the target sequence. The N-terminal formamide group of one molecule could be

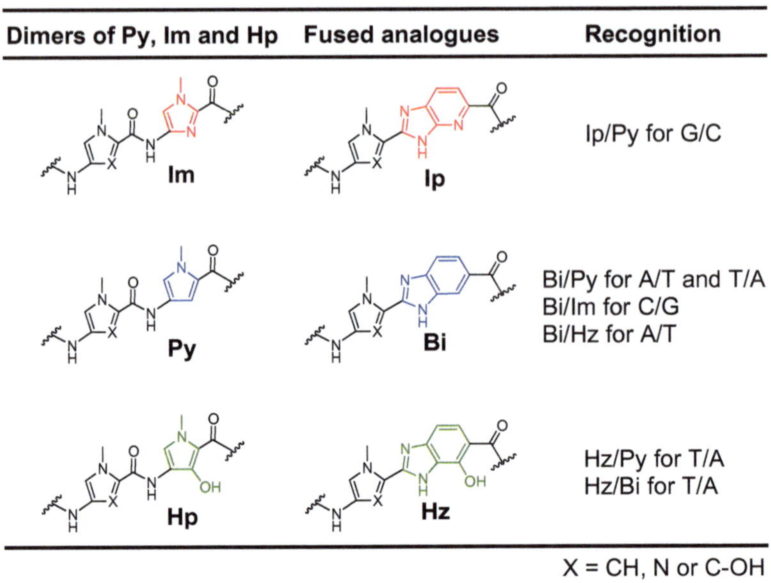

Fig. 1.7 Comparison of Py, Im and Hp dimers with the fused benzimidazole derivatives and their specificity of recognition. The image is quoted from Ref. [10i]. Copyright 2018 Elsevier

stacked with the C-terminal heterocycles of another molecule (Fig. 1.8a) [25a]. A subsequent study demonstrated that like Py/Py, the formamide/Py pair showed preference for A/T and T/A and that like Py/Im, the formamide/Im pair preferred C/G, but this selectivity for the C/G pair was lost when formamide/Im was placed next to a –PyPy–/–PyPy– pairing [25b]. Furthermore, fluorescent *p*-anisylbenzimidazole (Hx) and *p*-anisyl-4-azabenzimidazole (azaHx) attached to a Py–Im dimer unit and were found to function as recognition units mimicking formamide/Py (or consecutive Py–Py) and formamide-Im (or Py–Im), respectively (Fig. 1.8b) [25c, d].

1.3 Varied Structures of Py–Im Polyamides

A hairpin motif is most commonly used because of its versatility and ease of synthesis, but a variety of structures have been developed to enhance the affinity and specificity of Py–Im polyamides and to extend their target base pair range (Fig. 1.9).

Linear Py–Im polyamides are the most conventional motifs and usually bind as a 2:1 ligand–DNA complex. Both fully and partially overlapping binding in the forward binding orientation (N to C along the 5′–3′ direction of the DNA helix) have been reported, and these compounds were reported to target up to 16 bp, which is the length required for specific recognition of a unique site in the whole human genome (3 billion base pairs) [27]. In contrast, β-linked linear polyamides were found to bind

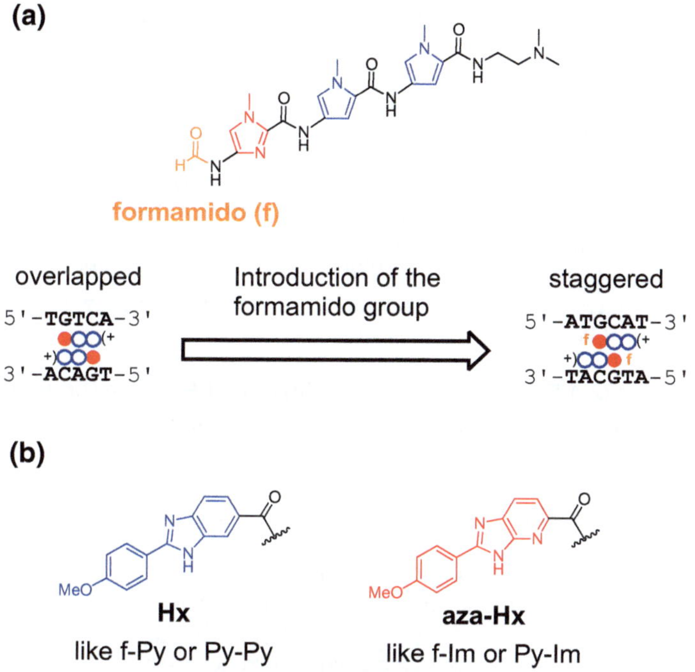

Fig. 1.8 **a** Structure and the staggered binding motif of the formamido-modified linear polyamide compared with the overlapping binding motif. **b** Structures and behaviors of Hx and aza-Hx. The image is quoted from Ref. [10i]. Copyright 2018 Elsevier

as 1:1 ligand–DNA complexes in the reverse orientation. In this binding mode, Im targeted G and C in preference to A and T, and Py and β preferentially bound A and T over G and C [28].

H-pin and U-pin polyamides include linked Py and Im [8, 29]. H-pins and U-pins connect two antiparallel oligomers via an aliphatic linker between the *N*-methyl groups of the central two Py residues and of the outer Py and Im, respectively. Linked Py and Im in U-pins could tolerate G/C base pairs at the turn moiety.

Because there is a possibility that γ-linked hairpin Py–Im polyamides can form at least two different conformations, i.e., hairpin and extended, which would result in different specificities, Dervan and coworkers developed cyclic Py–Im polyamides in which the C- and N-termini of the hairpin polyamides were covalently linked by a second γ-turn [12a, b, 30]. Cyclic polyamides exhibited higher binding affinity than the corresponding hairpin polyamides. A library of these compounds could be established using optimized solid-phase synthesis and the macrocyclization method [30f]. In addition, cyclic polyamides provided crystal structural data for DNA–polyamide complexes at atomic resolution (1.18 and 0.95 Å), which revealed that cyclic polyamides could function as allosteric modulators of DNA by perturbing the groove width and helix bending of DNA structures [12]. In contrast, our group

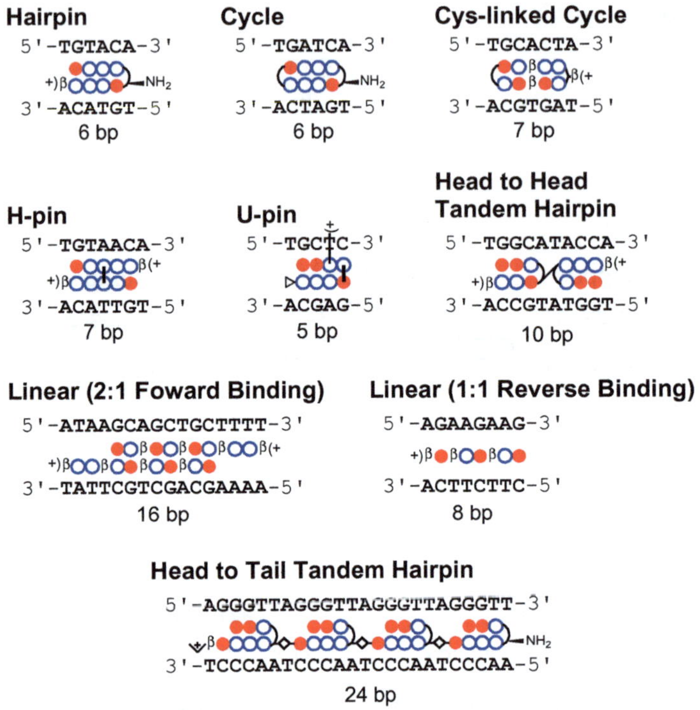

Fig. 1.9 Schematic representation of various Py–Im polyamide structures and their binding. The image is quoted from Ref. [10i]. Copyright 2018 Elsevier

developed different Cys-linked cyclic Py–Im polyamides by intramolecular coupling of the cysteine with N-terminus chloroacetyl residue [31]. These novel cyclic polyamides could target sequences longer than 6 bp, which is the upper limit for the cyclic polyamides reported by Dervan's group. Furthermore, we have developed new Py–Im polyamide dimers derived from these Cys-linked cyclic polyamides. These new dimerized polyamides could recognize up to 16 bp in 1:1 binding mode [32].

Connecting multiple hairpin moieties effectively expands their recognition sites. Dervan and coworkers demonstrated two types of tandem hairpin polyamides. The first was a head-to-head motif linking two γ-turn units [33]. They extended the linker length between the two hairpins and succeeded in targeting the two minor grooves on the nucleosome to stabilize the histone–DNA complex against dissociation [34a, b]. Using the head-to-head polyamides targeting two sites flanking the ends of a G-quadruplex-forming sequence, we successfully induced G-quadruplex formation [34c]. The other type was a head-to-tail motif linking the C-terminus tail with the γ-turn [35]. This motif could be used in other biological applications, such as the visualization of telomeres [36]. By introducing solid-phase synthesis of the building block corresponding to the three-ring Py and Im beside the N-terminal and modifying the γ-turn with an (*R*)-amino group at the α-position [36b, c], we have made the synthesis of tandem hairpin polyamides straightforward and have succeeded in targeting up to 24 bp in human telomeres. This is currently the record for the longest binding site of synthetic, nonnucleic-acid-based, sequence-specific DNA-binding molecules [36f].

1.4 Regulation of Gene Expression by Py–Im Polyamides

Py–Im polyamides are cell permeable, localize to the nucleus in living cells without requiring any delivery agent, and have binding affinities comparable to those of transcription factors [37]. Therefore, many efforts have been made to regulate gene expression in living cells by perturbing the DNA–transcription factor interaction using Py–Im polyamides. In this section, we present some examples of such studies.

The first report by Dervan, Gottesfeld and their colleagues demonstrated the targeting of an eight-ring hairpin polyamide to a specific region of the transcription factor TFIIIA binding site within the intragenic promoter element of the *Xenopus* 5S RNA gene, which inhibited its transcription in cultured *Xenopus* cells [38]. This transcriptional inhibition has been observed under various conditions. Hairpin polyamides targeting DNA sequences adjacent to the binding sites for transcription factors Ets-1, LEF-1, and TBP inhibited their binding, thereby impairing HIV-1 transcription and virus replication in isolated human peripheral blood lymphocytes, without detectable cytotoxicity [39]. This result suggests the possibility of clinical applications for Py–Im polyamides.

Laemmli and coworkers reported that 1:1 binding β-linked linear polyamides targeting *Drosophila melanogaster* A/T-rich satellite I and III- and GAGAA repetitive satellite V-mediated satellite-specific chromatin opening induced specific gain- and loss-of-function phenotypes [28a–c].

Vascular endothelial growth factor (VEGF) and its receptors play an important role in tumor angiogenesis and are upregulated by hypoxia. Hairpin polyamides designed to bind the hypoxia-response element (HRE) were found to inhibit the binding of hypoxia-inducible factor (HIF) to HRE in HeLa cells, resulting in a reduction in the expression of VEGF and a number of hypoxia-inducible genes (Fig. 1.10a) [40]. Further studies using a tumor xenograft model demonstrated that polyamides targeting HRE reduced the expression of proangiogenic and prometastatic factors, inhibited the formation of new tumor blood vessels, and suppressed tumor growth [40d]. Furthermore, the polyamides targeting HRE were also effective in chemotherapy-resistant multiple myeloma cells, in which hypoxic conditions are implicated in resistance development [40e].

Py–Im polyamides targeting the androgen-response element (ARE) in the PSA promoter of LNCaP and VCaP cells disrupted the androgen receptor (AR)–ARE interaction and inhibited expression of androgen-induced genes (Fig. 1.10b) [30e, 41a–b]. Polyamide treatment resulted in depletion of the RNAP2 large subunit RPB1 and accumulation of p53, without DNA damage, thereby inducing antitumor activity in a prostate tumor xenograft model [41c]. Furthermore, the polyamides could also

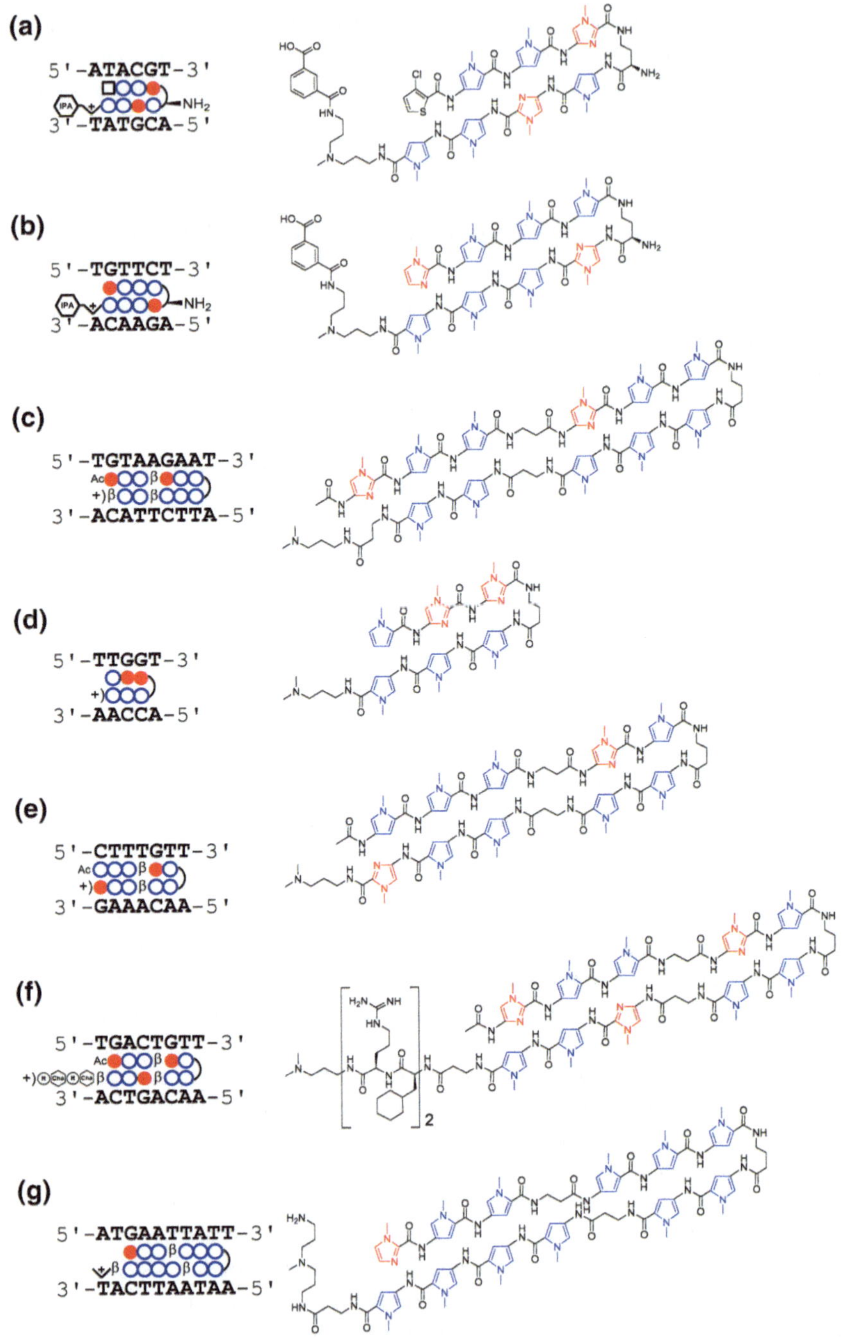

Fig. 1.10 Schematic representation and structures of hairpin Py–Im polyamides for **a** HRE, **b** ARE and GRE, **c** AP-1 site in the hTGF-β1 promoter, **d** NF-Y site in the *Topo IIα* promoter, **e** SOX2 binding site, **f** TFAM in mitochondria, and **g** AT rich regions of HPV genome. The image is quoted from Ref. [10i]. Copyright 2018 Elsevier

bind the glucocorticoid-response element (GRE) to inhibit the glucocorticoid receptor (GR)–DNA interaction, resulting in effects on a set of glucocorticoid-induced and -repressed genes [41d]. By interfering with both AR- and GR-driven gene expression, the polyamides were effective in LREX′ prostate cancer models resistant to enzalutamide, a potent AR ligand-binding domain antagonist [41e]. However, the polyamides targeting ARE were found to cause replication stress, resulting in an accumulation of S-phase cells [41f]. This is believed to be one possible mechanism of their anticancer effect.

Dervan, Gottesfeld and coworkers reported β-linked linear polyamides targeting GAA/TTC triplet repeats (Fig. 1.9) [28f, h, i]. Hyperexpansion of GAA/TTC repeats in the first intron of a nuclear gene encoding the essential mitochondrial protein frataxin was found in the neurodegenerative disease, Friedreich's ataxia (FRDA). The expanded repeats adopt a non-B-DNA structure and repress the expression of the frataxin gene (*FXN*). The linear polyamides targeting GAA/TTC repeats locked the repeats into double-stranded canonical B-DNA formation and increased *FXN* expression in lymphoid cell lines from a FRDA patient [28f]. They also impeded the expansion of the repeats by binding and displacing mismatch repair enzyme MSH2 from the repeats in induced pluripotent stem cells (iPSCs) from FRDA patients [28h].

Fukuda, Matsuda and their colleagues reported hairpin Py–Im polyamides targeting an AP-1 transcription factor binding site in the transforming growth factor-β1 (TGF-β1) promoter, which inhibited the expression of the gene (Fig. 1.10c) [42]. Subsequent studies showed that the polyamides could inhibit expression of TGF-β1 in the renal cortex and mitigate the increase in urinary protein and albumin in Dahl-S rats [42b]. The polyamides reduced glomerulosclerosis and interstitial fibrosis without side effects [42c]. These polyamides targeting the TGF-β1 promoter are potential gene silencers for use in treatment of renal diseases. In other contexts, the polyamides were able to suppress neointimal hyperplasia in injured rat carotid artery by inhibiting TGF-β1 and connective tissue growth factor (CTGF) expression and reducing the extracellular matrix [42d]. In a rat model of corneal alkali burns, the polyamides inhibited TGF-β1 expression in the burned cornea, improved corneal tissue wound healing, reduced corneal stromal scarring and prevented corneal haze formation [42e]. In addition, the polyamides were found to inhibit hypertrophic scarring in rats and marmosets and could be a useful and practical treatment for patients with hypertrophic scarring after surgical operations or skin burns [42f, g].

Hochhauser, Hartley and coworkers reported the development of Py–Im polyamides designed to block the repressive binding of the transcription factor NF-Y within the Topoisomerase IIα (*Topo IIα*) promoter (Fig. 1.10d) [43a]. Confluent NIH3T3 cells expressing lower levels of *Topo IIα* are resistant to etoposide treatment, but the polyamides inhibited repression of *Topo IIα* and enhanced etoposide-induced DNA double-strand breaks in these cells, suggesting that the polyamide could modulate not only gene expression but also the chemosensitivity of cancer cells [43].

Py–Im polyamides can also be used to control cell differentiation. Our group reported the development of a polyamide that could inhibit the SOX2 protein–DNA interaction and caused mesoderm induction in human iPSCs (Fig. 1.10e). This mesoderm induction by the polyamide resulted in the successful generation of contracting cardiomyocytes after cells were treated with a Wnt/β-catenin signaling inhibitor [44].

Py–Im polyamides penetrate the cellular membrane and localize in the nucleus without the requirement for additional delivery agents, but polyamides modified with mitochondria-penetrating peptide (MPP) composed of cyclohexylalanine (Cha) and D-arginine (Arg) showed specific localization inside mitochondria in HeLa cells (Fig. 1.10f). The modified polyamides could recognize a specific DNA sequence and inhibit the binding of mitochondrial transcription factor A (TFAM) to DNA, resulting in the repression of the *ND6* gene [45]. These polyamide-MPP conjugates provide the basis for designing DNA-binding molecules that can affect the mitochondrial genome in a sequence-specific manner.

Polyamide-treatment can reduce the episome level of viruses such as human papillomavirus (HPV). Parekh and colleagues reported a tandem hairpin polyamide that could interfere with the binding of viral E2 protein to DNA in vitro [35c]. Recently, Bashkin and coworkers demonstrated hairpin polyamides targeting A/T-rich regions of the HPV genome that would be good targets for polyamides because they are in or near the binding sites of viral E1 or E2 proteins; therefore, polyamides targeting these A/T-rich regions would interfere with viral maintenance. In host cells, these hairpin polyamides targeting A/T-rich regions exhibited antiviral activity against high-risk HPV strains 16, 18, and 31 without detectable cellular toxicity (Fig. 1.10g) [46]. These polyamides will be applied as novel antiviral therapeutics.

1.5 Repression of Gene Expression by Sequence-Specific DNA Alkylation

As described above, Py–Im polyamides can regulate gene expression by disrupting DNA–protein interactions, especially within promoter regions. However, it is difficult to regulate gene expression when the Py–Im polyamides are bound to the coding region, because it is difficult to stop RNA polymerase (RNAP) transcription using noncovalently bound polyamides. Py–Im polyamides conjugated to alkylating agents could covalently bind to specific sequences and inhibit transcription, resulting in gene silencing. Indeed, polyamides conjugated to an alkylating moiety and bound to coding regions were able to inhibit T7 RNAP and generate truncated mRNA in vitro [47].

Py–Im polyamides conjugated to either the nitrogen mustard chlorambucil (Chl) [47b, 48], or *seco*-1-chloromethyl-5-hydroxy-1,2-dihydro-3*H*-benz[*e*]indole (CBI) [49, 50], an analogue of the natural product (+)-CC-1065, have been reported (Fig. 1.11). The former mediated sequence-specific alkylation of adenine N3 and guanine N7 and the latter alkylated adenine N3. Chl conjugates both alkylated and crosslinked DNA [48b].

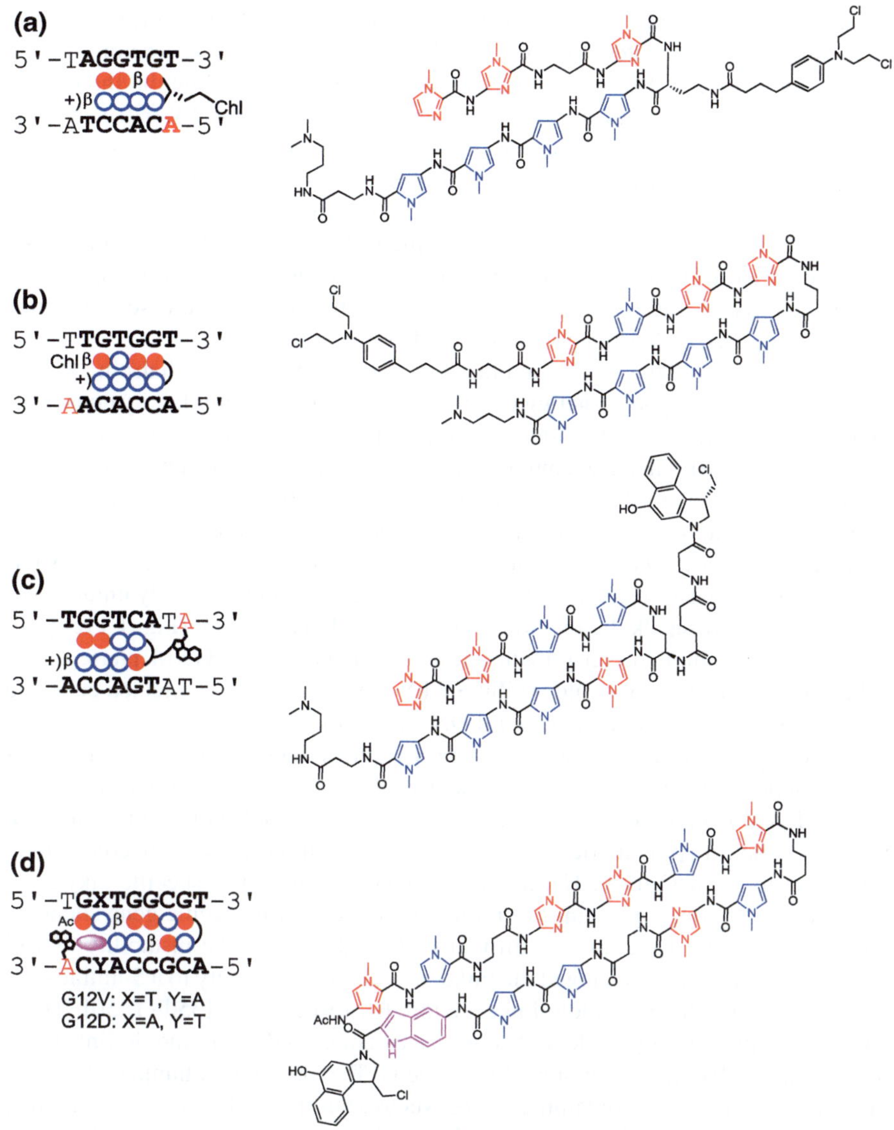

Fig. 1.11 Schematic representation and structures of hairpin Py–Im polyamides-alkylation agent conjugates. **a** α-DABA hairpin polyamide-Chl conjugate for histone H4c gene coding region. **b** Polyamide-Chl conjugate for RUNX family binding sites. **c** Polyamide-*seco*-CBI conjugate connected at γ-turn. **d** Polyamide-*seco*-CBI conjugate linked by indole linker at C-terminus for KRAS codon mutation. Bases recognized by polyamide moieties are represented as bold and alkylated bases identified by in vitro PAGE analysis are represented as red. The image is quoted from Ref. [10i]. Copyright 2018 Elsevier

Dervan, Gottesfeld and coworkers demonstrated that hairpin Py–Im polyamide–Chl conjugates that could bind within the coding region of the histone H4c gene could downregulate expression of the gene and inhibit the growth of several cancer cell lines in vitro, and cancers in the murine SW620 xenograft model of human colon carcinoma and the murine K562 xenograft model of chronic myelogenous leukemia [48c–h]. The Chl moiety was attached to the γ-turn, but surprisingly, substituting γ-diaminobutyric acid (DABA) with α-DABA decreased the rate of alkylation but increased its specificity (Fig. 1.11a). Furthermore, mice treated with an α-DABA hairpin polyamide–Chl conjugate showed no significant toxicity compared with control mice, although lethal effects were seen in mice treated with γ-DABA hairpin polyamide–Chl conjugate [48e].

Our group and that of Kamikubo recently reported a hairpin Py–Im polyamide–Chl conjugate targeting the RUNX transcription factor family consensus binding site in which the Chl residue was attached to the N-terminus of the polyamide moiety (Fig. 1.11b). This new conjugate induced p53 pathway-dependent cell death and was effective against acute myeloid leukemia (AML) cells and against poor-prognosis solid tumors in a xenograft mouse model of AML [48i].

Sequence-specific alkylation using Py–Im polyamide-*seco*–CBI conjugates has also been investigated. Dervan and coworkers developed a hairpin polyamide-*seco*-CBI conjugate whose alkylation moiety was attached to the γ-turn (Fig. 1.11c) [49]. This conjugate inhibited cell growth and arrested cells in the G2/M phase of the cell cycle. Furthermore, it could inhibit SV40 virus replication in infected cells and penetrate SV40 virions to damage viral DNA [49b, c].

Previously our group had used a cyclopropylpyrroloindole (CPI) alkylating moiety, segment A of duocarmycin 86, linked to the C-terminus of a hairpin Py–Im polyamide via a vinyl linker [47a, 51a–d]. CPI alkylates the N3 of adenine and guanine. Using hairpin polyamide–CPI conjugates, we demonstrated a selective reduction in luciferase activity in HeLa cells mediated by sequence-specific alkylation of the coding region, suggesting the possibility of sequence-specific gene silencing [51d]. However, to overcome the difficulty of synthesis and the instability of CPI having a vinyl linker, we changed the linker and alkylation moiety to a 5-amino-1*H*-indole-2-carbonyl linker (indole linker) and CBI, respectively [35d–f, 50]. These new hairpin Py–Im polyamide–CBI conjugates connected by an indole linker that targeted the GFP coding region could silence GFP expression in human colon carcinoma cells (HCT116) containing a GFP vector, but the CBI conjugates targeting the promoter and noncoding regions could not silence gene expression [50b]. However, they exhibited antitumor activity against several cancer cell lines and in mouse models [50c–e].

Our group and that of Nagase also developed a hairpin Py–Im polyamide–CBI conjugate targeting oncogenic KRAS codon-12 mutants (Fig. 1.11d). Many colon cancer patients have the KRAS codon G12D or G12 V mutations. In human colon cancer-derived cells, this conjugate could suppress mutant KRAS expression and induce p53-dependent cellular senescence and DNA damage, resulting in apoptosis. In addition, this polyamide could suppress tumor growth in a xenografted nude mouse model [50f–h]. This newly developed approach of directly targeting specific sequences using Py–Im polyamide–alkylating moiety conjugates will be applicable to disease therapeutics.

1.6 Transcriptional Activation

Py–Im polyamides have also been investigated as artificial transcriptional activators. Dervan and coworkers demonstrated that the DNA-binding domain of Py–Im polyamides could be tethered to short peptides such as VP16 or to the activation domain of nonpeptide small molecules [52a–d]. Recruitment of transcription factors to DNA by these Py–Im polyamide–protein-binding peptide conjugates could be confirmed by gel shift assay [52e–i]. Using these polyamide-derived artificial activators, Dervan and colleagues succeeded in inducing transcription in a sequence-specific manner in an in vitro system [52a–d]. In 2007, Kodadek and coworkers reported the first example of a polyamide-based, cell-permeable, synthetic transcriptional activator that could activate the transcription of a reporter gene in HeLa cells. The compound was composed of a DNA-binding Py–Im polyamide fused with a peptoid, which was isolated from a combinatorial library and bound to the KIX domain of the mammalian transcription coactivator CREB-binding protein (CBP) (Fig. 1.12a). This synthetic activator increased luciferase expression in HeLa cells transfected with a luciferase reporter plasmid containing six polyamide-binding sites in its promoter. Furthermore, expression of endogenous chromosomal genes containing multiple polyamide binding sites within their promoters could be upregulated [53]. These results validated the applicability of synthetic Py–Im polyamides to gene activation.

We used a different approach to achieve gene activation by focusing on the histone deacetylases (HDACs) and histone acetyltransferases (HATs) that govern histone acetylation and DNA accessibility to transcription factors [54–56] We conjugated hairpin Py–Im polyamides to suberoylanilide hydroxamic acid (SAHA), known to be a HDAC inhibitor (Fig. 1.12b) [54]. Our first reported conjugate, designed to target the sequence present in the promoter region of the tumor suppressor *p16* gene, induced histone H3 lysine 9 acetylation of the *p16* promoter and morphological changes in HeLa cells [54a]. In a subsequent study, we established a library of SAHA–hairpin Py–Im polyamide conjugates targeting 6 bp sequences, and evaluated their effect on gene expression in mouse embryonic fibroblasts (MEFs) or human dermal fibroblasts (HDFs) [54b–i]. Screening studies in MEFs identified those conjugates that could induce endogenous transcription of reprogramming factors (*Oct-3/4*, *Nanog*, *Sox2*, *Klf4*, and *c-Myc*) and multiple pluripotency genes, including *Rex1* and *Cdh1* [54b–d]. Microarray and RT-qPCR studies demonstrated the remarkable ability of different SAHA–polyamide conjugates to activate the transcription of specific clusters of genes and noncoding RNAs [54e]. Functional analysis identified four different SAHA–polyamide conjugates: K, I, X and L (Fig. 1.12b) [54f–i]. In HDFs, conjugates K, I, and X induced transcriptional activation of germ-cell-specific and PIWI pathway genes, *OCT-3/4*-regulated pluripotent pathway genes,

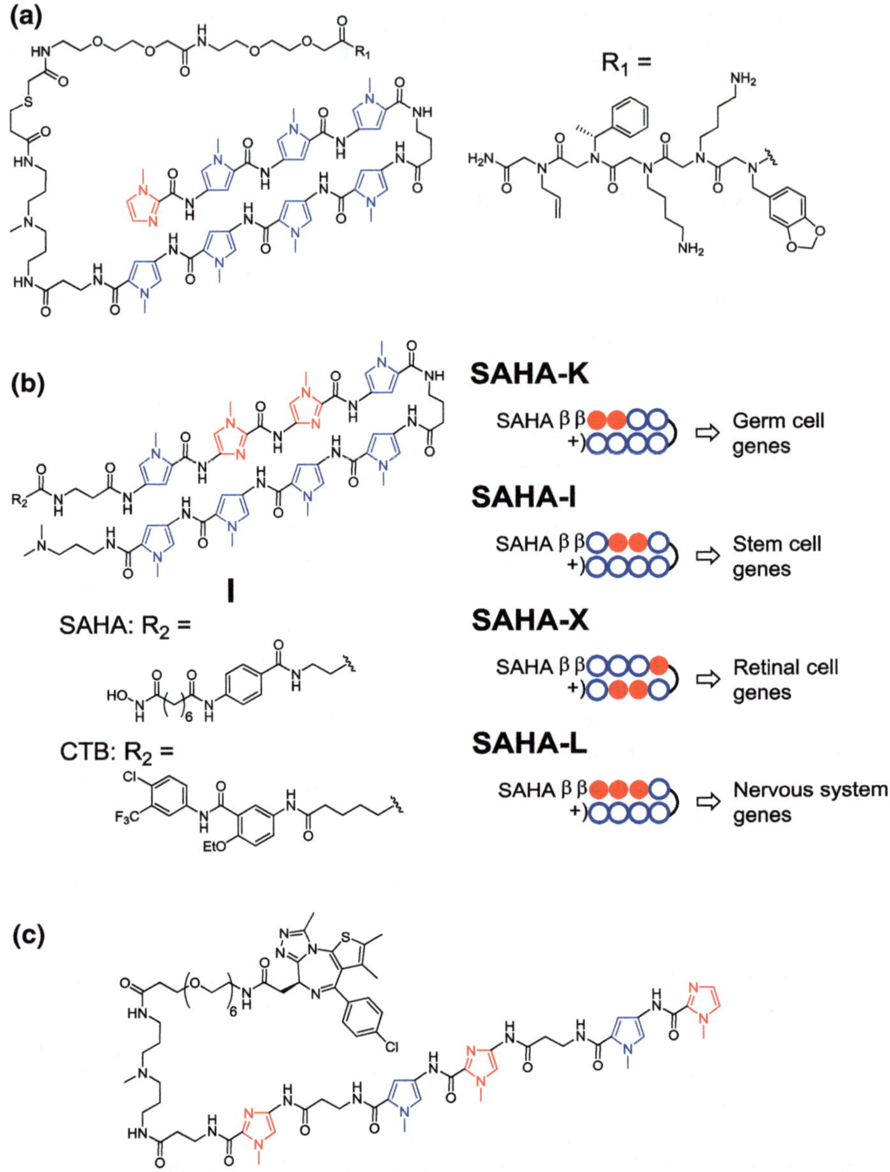

Fig. 1.12 Py–Im polyamide conjugates for transcriptional activation. **a** Structure of a polyamide–KIX domain-binding peptoid conjugate. **b** Structure of a polyamide–SAHA or –CTB conjugate and the schematic for characterized SAHA–polyamide conjugates. **c** Structure of a polyamide–BRD4 protein ligand JQ1 conjugate targeting GAA/TTC repeats. The image is quoted from Ref. [10i]. Copyright 2018 Elsevier

and retinal-tissue-related genes, respectively [54f–h]. Conjugate L could activate genes associated with brain synapses and neurotransmitter networks in BJ human foreskin fibroblasts and 201B7-iPS cells. Furthermore, the long-term incubation of 201B7-iPS cells with conjugate L induced morphological changes and promoted a neural progenitor phenotype [54i].

We also considered the regulation of HATs and investigated the possibility of conjugating a HAT activator *N*-(4-chloro-3-trifluoromethyl-phenyl)-2-ethoxy-benzamide (CTB) to a hairpin Py–Im polyamide. Interestingly, CTB fused with DNA-binding domain I activated a cluster of genes that was similar to that activated in HDFs by SAHA-conjugate I (Fig. 1.12b) [55]. This study confirmed the interchangeable functions of HDACs and HATs in gene regulation. However, a hairpin Py–Im polyamide coupled with HAT inhibitor C646 exhibited antiproliferative activities in cancer cell lines via upregulation of p53 target genes [56].

Ansari and coworkers have focused on BRD4 protein, which binds acetylated lysine on histones and recruits positive transcription elongation factor b (P-TEFb) to facilitate transcription elongation. Using a Py–Im polyamide targeting GAA/TTC repeats relevant to FRDA conjugated with BRD4 protein ligand JQ1 (Fig. 1.12c), they successfully recruited BRD4 to the target site and induced endogenous *FXN* expression in FRDA patient cells, where the *FXN* gene is repressed [57]. This result suggested that synthetic transcriptional activators constructed from Py–Im polyamides are applicable to clinical research and treatment.

Recently, our group and that of Lee established the NanoScript platform (Fig. 1.13) [58]. A NanoScript is a nanoparticle, such as a gold nanoparticle, that is linked to a nuclear localization domain, an activation or repression domain, and a DNA-binding domain, and which mimics the structure and gene-regulating activity of transcriptional factors. The nuclear localization and activation domains are peptides and the DNA-binding domain is a programmable hairpin Py–Im polyamide. NanoScripts could penetrate the plasma membrane and localize within the nucleus, where they could target endogenous genes without a requirement for a viral vector [58a]. NanoScript has been used mainly for induction of stem cell differentiation. For example, a NanoScript designed to replicate the actions of natural transcription factors specific for myogenic regulatory factors (MRFs) interacted with DNA to activate the four critical genes that regulate myogenesis, *MYOD*, *MYOG*, *MYF5*, and *MYRF* and induced differentiation of stem cells into muscle cells [58b]. Another NanoScript designed to target and activate the *SOX9* gene, which was modified with a HAT-activating molecule CTB, could enhance expression of the gene in mesenchymal stem cells and induce chondrogenic differentiation [58c]. A NanoScript could also mimic the fundamental structure and gene-silencing function of repressor transcriptional factors if a peptidic repressor domain was substituted for the activator domain. The repressing NanoScript designed to target a sequence within the *SOX9* promoter could repress its expression in human neural stem cells, resulting in enhancement of neuronal differentiation [58d]. This NanoScript platform will provide scientists with a new tool for gene regulation.

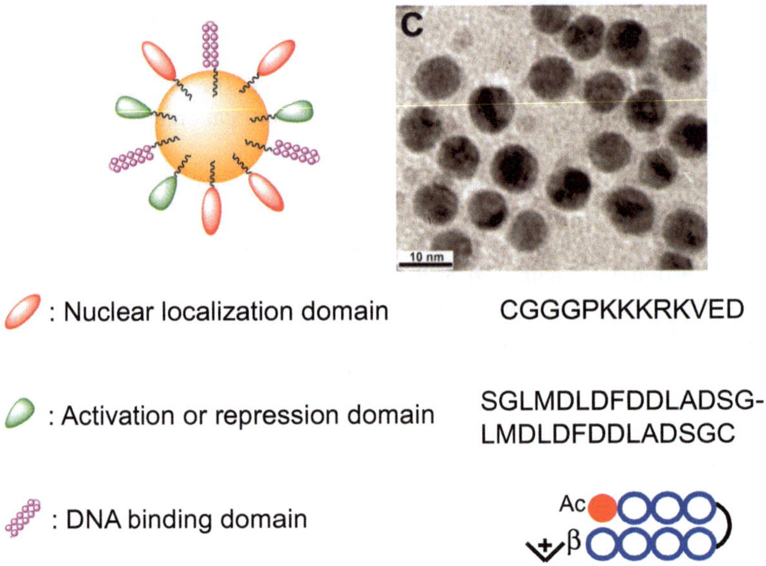

Fig. 1.13 Schematic representation of NanoScript including nuclear localization domain, either activation or repression domain, and DNA binding domain attached to a nanoparticle, and example of each domain [58a], and a TEM microscopic image of NanoScript. The image is quoted from Refs. [10i, 58a]. Copyright 2018 Elsevier and Copyright 2014 American Chemical Society

1.7 Py–Im Polyamides for Imaging

Py–Im polyamides conjugated to imaging agents such as fluorescent compounds have been used mainly for detection of specific sequences, discrimination of mismatched sequences, and investigation of the localization of polyamides [59]. In this section, we review some examples of the use of polyamides and their conjugates for imaging.

In 2001, Laemmli and coworkers reported specific visualization of insect (5′-$(TTAGG)_n$-3′ repeats) and vertebrate (5′-$(TTAGGG)_n$-3′ repeats) telomeres in fixed, permeabilized cells and in chromosomal cell spreads using Texas Red-labeled tandem hairpin Py–Im polyamides targeting the above repetitive sequences. A widely used fluorescence in situ hybridization method using peptide nucleic acid probes requires harsh treatment, which is time consuming and carries the risk of destroying cellular structures, but this new method does not require such denaturation. Furthermore, Laemmli et al. succeeded in visualizing insect telomeres in living cells [36a]. Recently, we collaborated with Maeshima, Nokihara and coworkers to establish an easy, straightforward, solid-phase method for synthesis of tandem hairpin Py–Im polyamides targeting vertebrate telomeric repetitive sequences, and succeeded in synthesizing tandem dimer, trimer, and tetramer polyamides recognizing 12 bp, 18 bp, and 24 bp, respectively. The specificity of visualization using tetramethylrhodamine (TAMRA)-labeled polyamide probes was improved when the number of target base pairs was increased and the hinge region between hairpin moieties was optimized (Fig. 1.14) [36b–f]. We have demonstrated successfully that these polyamide probes could visualize telomeres in mouse and human tissue sections and could be applied to quantitative measurements of telomere length with single-cell resolution [36e]. We are currently trying to extend the applicability of these polyamide probes for telomere visualization.

Other than telomere-targeting fluorescent polyamides, it has been reported that fluorescently labeled polyamides targeting the TTCCA-motif repeated in the heterochromatic regions of chromosomes 1, 9, and Y in human cells [60], and polyamides recognizing GAGAA repeats in satellite V or AT-rich regions in satellite I and III of *Drosophila melanogaster* [28a–c] could stain specific regions of chromosomes without requiring harsh denaturation. Recently, Boutorine and coworkers reported a fluorescein isothiocyanate (FITC)-labeled polyamide that could visualize centromeres in fixed and living murine cells [61]. Further investigation and optimization of Py–Im polyamide probes is required to improve staining of specific sites in cells.

Subsequent studies have investigated the change in the fluorescent characteristics of Py–Im polyamide–dye conjugates after binding to matched or mismatched sequences. Dervan and coworkers designed and tested hairpin Py–Im polyamides linked to TAMRA at the *N*-methyl group of the central Py residue (Fig. 1.15a). Their fluorescence was quenched in the absence of dsDNA, likely because an intramolecular ground-state complex formed between TAMRA and the polyamide in the absence of DNA, and this facilitated electron transfer from the Py moieties to the excited-state TAMRA. Addition of dsDNA including a matching sequence generated at least 10-fold increase in fluorescence, but the increase in fluorescence was less when mismatched DNA was added [62]. Similar sequence-dependent enhancement was observed using perylene-conjugated polyamides for the detection of telomeric repeats [63]. To increase the fluorescence enhancement, Dervan and coworkers conjugated the C-terminus of Py–Im polyamides to a fluorescent DNA intercalator, thiazole orange (TO) (Fig. 1.15b). These polyamide–TO conjugates displayed a substantial increase in fluorescence (>1000-fold) after sequence-dependent binding [64]. Interestingly, hairpin polyamides incorporating Bi and its derivatives Ip and Hz fused with five-membered rings were found to show fluorescence enhancement in a sequence-specific manner without conjugation to any other fluorescent groups (Fig. 1.15c) [24e]. In contrast, our group developed polyamides conjugated to two pyrene moieties at the *N*-methyl group of Py residues. These novel fluorescent conjugates exhibited fluorescence enhancement via formation of excimers between two pyrene moieties upon their binding to DNA (Fig. 1.15d) [65]. This system of fluorescence enhancement upon binding to a matched sequence would be applicable to the identification of single-nucleotide polymorphisms (SNP) and visualization of specific sequences in cells with a high signal-to-noise ratio.

Py–Im polyamide–fluorescent dye conjugates have also been used for evaluation of polyamide localization in cells and mouse models. Several studies using such conjugates have demonstrated that Py–Im polyamides can localize to the nucleus in cells (Fig. 1.16a) [37, 42a, 66], but that the efficacy of localization depends on the cell type, the structure of the polyamide backbone, the fluorophore, the molecular size and

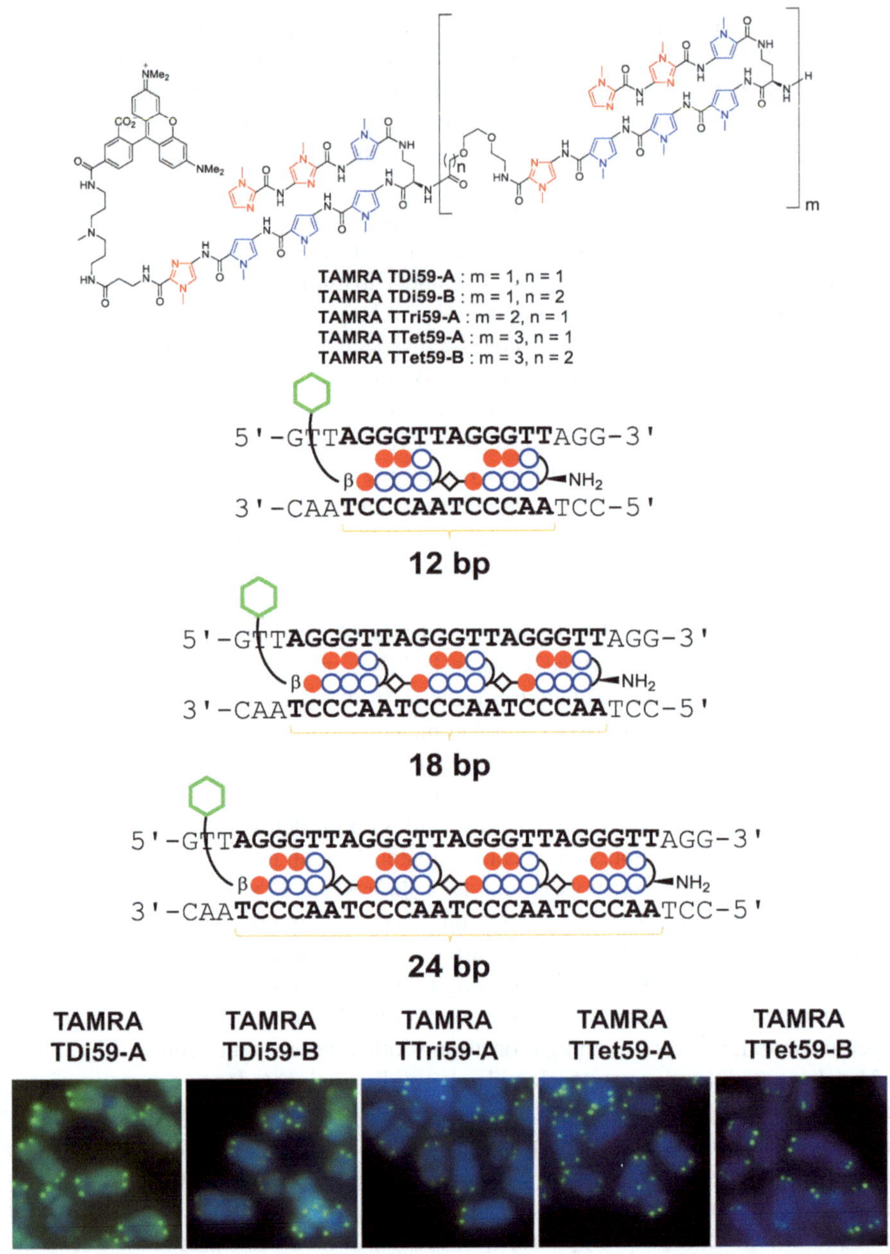

Fig. 1.14 Structures and schematic representations of tandem dimer, trimer and tetramer Py–Im polyamide probes labeled with TAMRA targeting telomeric repeats and comparative imaging of telomeres. The image is quoted from Ref. [10i, 36f] and modified. Copyright 2018 Elsevier and Copyright 2016 American Chemical Society

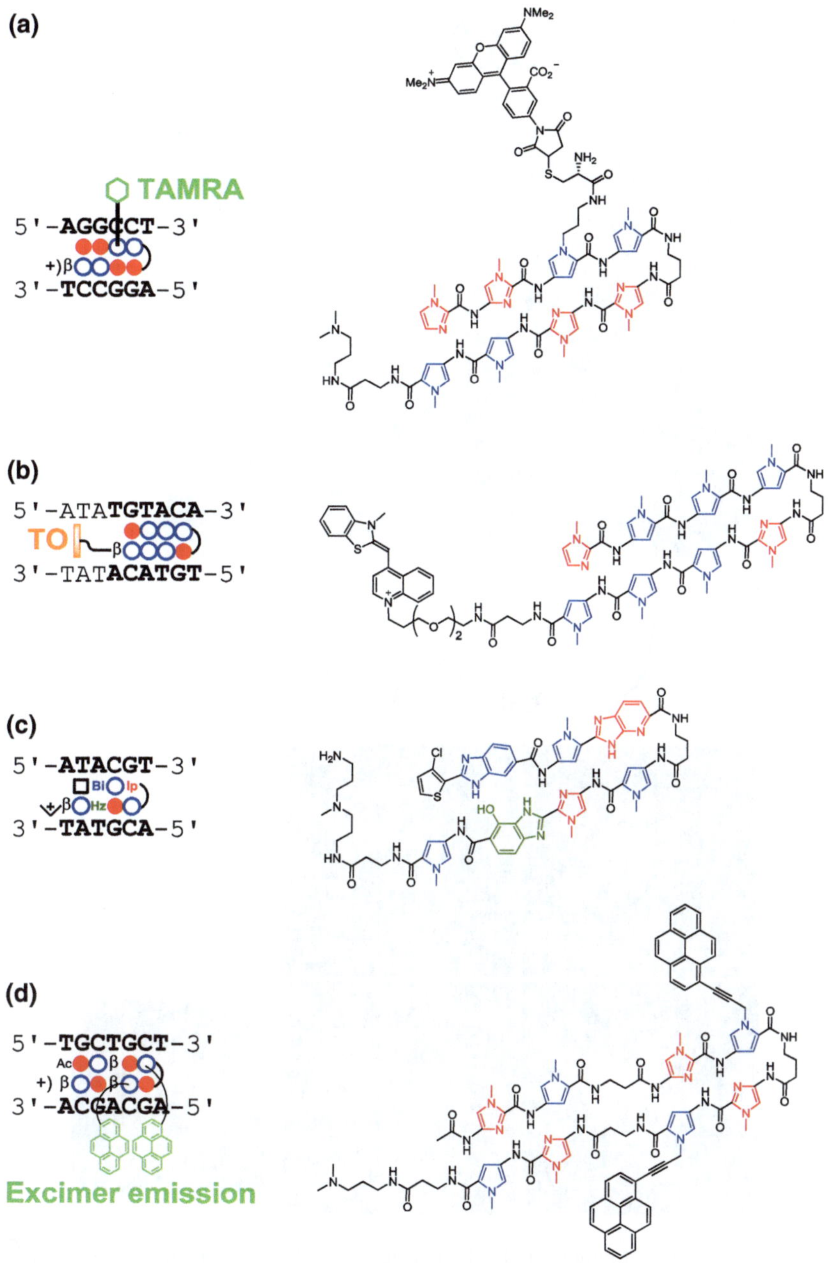

Fig. 1.15 Structures and schematic representations of fluorescent Py–Im polyamides that exhibit sequence-dependent fluorescence enhancement. **a** Polyamide conjugated to TAMRA at the *N*-methyl group of the Py residue. **b** Polyamide conjugated to the DNA intercalator TO. **c** Polyamides incorporating benzimidazole and its derivatives. **d** Polyamide conjugated to two pyrene residues forming an excimer. The image is quoted from Ref. [10i]. Copyright 2018 Elsevier

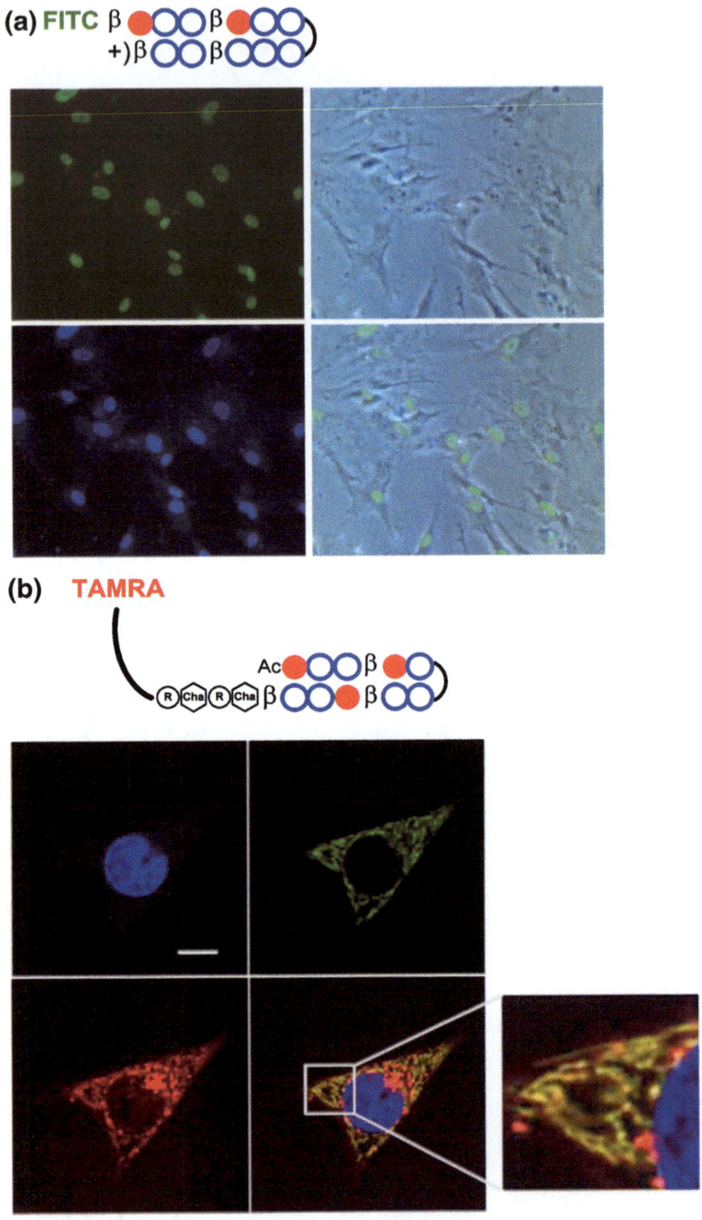

Fig. 1.16 Localization of fluorescent Py–Im polyamides within cells. **a** Nuclear localization of a FITC-labeled polyamide (green) targeting an AP-1 transcription factor binding site in the TGF-β1 promoter. Nuclear staining was performed using Hoechst 33342 (blue). **b** Mitochondrial localization of TAMRA-labeled polyamide incorporating a mitochondria-penetrating peptide (red). Nuclei and mitochondria were visualized by Hoechst 33342 (blue) and CellLight Mitochondria-GFP, BacMam 2.0 (green), respectively. The image is quoted from Ref. [10i]. Copyright 2018 Elsevier. Two pictures are reproduced from Refs. [42a, 45], respectively. Copyright 2005 The American Society for Pharmacology and Experimental Therapeutics and 2017 American Chemical Society

the presence of Im residues [37, 66]. Subsequent studies have attempted to improve the efficacy of localization in cells by structural modifications and improvements in solubility, such as modification of C-terminus by isophtalic acid and introduction of aryl group to the γ-turn, and improvements in solubility (discussed below) [40b, 67]. The dye conjugates could also be used for confirming the specific localization of polyamide-MPP to mitochondria (Fig. 1.16b) [45].

Radiolabeled Py–Im polyamides can enable in vivo observation of their uptake and provide information about their biodistribution and metabolism [18]. F-radiolabeled Py–Im polyamides administered intravenously to C57 mice were evaluated using positron emission tomography/computed tomography (PET/CT). The labeled Py–Im polyamides localized rapidly to the liver immediately after injection and were cleared from the liver by GI-mediated excretion, the speed of which was dependent on the structure of the polyamide (Fig. 1.17a) [68a]. However, because ^{18}F is a short-lived radionuclide ($t_{1/2}$ = 109.8 min), the ^{18}F-labeled polyamides could be traced for only 2 h following injection. To obtain more extended monitoring to investigate the chemotherapeutic potential of the polyamides, they were labeled with ^{14}C. Py–Im polyamides administered subcutaneously to mice bearing a tumor xenograft were found to localize to the xenograft (Fig. 1.17b) [68b, c].

Fluorescently labeled Py–Im polyamides can also be used for other applications. Burley and coworkers established a photonic wire model using Pacific Blue (PB) as the initial donor chromophore, which was covalently attached to one terminal of DNA, and Cy3 as the terminal energy acceptor chromophore attached to the other terminal, and a Py–Im polyamide–dye conjugate that could regulate the binding site within DNA. The polyamide conjugates functioned as energy transporters and enhanced the efficiency of Forster resonance energy transfer (FRET) between the separated donor and acceptor along the dsDNA track. This approach allowed the observation of FRET over 27 nm (80 bp) [69]. In addition, we and the Harada group successfully observed FRET between Cy3- and Cy5-labeled polyamides on the nucleosome [70]. This result opens up the possibility of characterizing dynamic interactions within protein–DNA complexes using fluorescently-labeled polyamides.

Pyrene can inject electrons into DNA and the subsequent electron transfer along DNA can be detected by ^{Br}U. We developed Py–Im polyamide–pyrene conjugates as sequence-specific electron injectors. After UV irradiation and subsequent uracil DNA glycosylase treatment and heating, electrophoretic analysis demonstrated that DNA was cleaved at the ^{Br}U when the electron was injected by sequence-specific polyamide conjugates. This platform would be useful for identifying the binding sites of Py–Im polyamides [71].

1.8 Other Functions and Applications of Py–Im Polyamides

We have reviewed the functions and applications of Py–Im polyamides and their conjugates, focusing on the frequently studied areas of gene regulation and fluorescence. In this section, we would like to review other possible functions and applications.

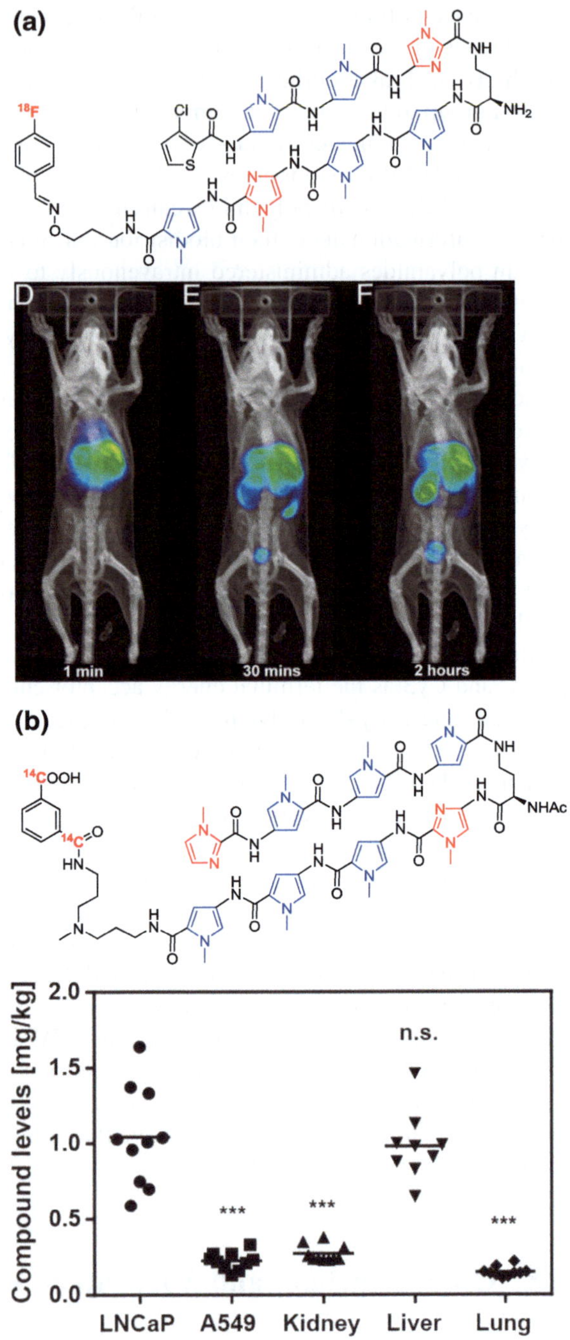

Fig. 1.17 Structures of radiolabeled Py–Im polyamides and data obtained after their injection. Chemical structures are quoted from Ref. [10i]. Copyright 2018 Elsevier. **a** ^{18}F-labeled polyamide and PET/CT images of mice. These images are modified from those appearing in Ref. [68a]. Copyright 2008 National Academy of Sciences. **b** Levels of ^{14}C-labeled polyamide targeting ARE in LNCaP (prostate) and A549 (lung) tumor xenografts, compared to those in major host organs (kidney, liver, lung). This graph is reproduced from Ref. [68c]. Copyright 2014 American Chemical Society

Sequence-specific binding of Py–Im polyamides can be applied to DNA nanotechnology [72]. Dervan and coworkers reported that biotin-labeled polyamides could recruit and localize streptavidin to designated positions on two-dimensional DNA tiles [72a, b]. Our group reported that using a DNA origami scaffold carrying five different dsDNA sequences in its cavities, the biotin-labeled polyamide-*seco*-CBI conjugate could bind and alkylate dsDNA in a sequence-specific manner and recruit streptavidin to the cavity that included its target sequence (Fig. 1.18a) [72c]. Heckel and coworkers demonstrated that polyamides could connect two DNA rings and successfully produced a DNA catenane (Fig. 1.18b) [72d–f].

Sequence-specific binding of Py–Im polyamides can also be used to cleave DNA, mainly by conjugating them to a DNA-cleaving agent. Redox-active metal complexes mediate oxidative DNA damage [73, 74]. Therefore, polyamide–EDTA–Fe(II) conjugates could provide information about sequence-specific binding of the polyamides by inducing sequence-specific DNA cleavage (Fig. 1.19) [6b, 73]. In contrast, a polyamide–bicyclic enediyne conjugate and a polyamide–bis(guanidinium)alcohol conjugate were also reported to cleave DNA in the absence of metal ions [75]. Furthermore, recruitment of Topoisomerase I and cleavage of DNA in a sequence-specific manner have been achieved by conjugating polyamides to camptothecin [76].

Posttranslational modification of histones plays an important role in gene regulation and a method for modulating histone modification would be useful to elucidate the mechanisms of gene regulation. Kanai and coworkers established a method for modifying histones in vitro using synthetic nucleosome binders conjugated to synthetic catalysts without relying on enzymatic activity [77]. They conjugated a hairpin Py–Im polyamide or histone H2A-H2B binding peptide to the synthetic acyl transfer catalyst DSH to produce a site-selective histone modification system. The conjugates bound to specific regions of nucleosomes and successfully introduced an acyl group from an acyl donor, such as acetyl-CoA, to lysines close to the catalyst moiety of the histone in a region-specific manner (Fig. 1.20a). Using this method, Kanai and coworkers promoted histone H2B lysine-120 acylation and found that this modification modulated high-order chromatin structures by reducing internucleosomal interactions [77a]. Further, a combination of an arginine-conjugated acyl donor (nucleosome-binding acyl donor) with a Py residue-conjugated catalyst (nucleosome-binding catalyst) succeeded in preferentially acylating lysines on histone tails and modulating chromatin structure in a way similar to HAT (Fig. 1.20b). Using this method, they demonstrated a reduction in self-association of nucleosomes and induction of transcription [77b]. These synthetic nucleosome-modification systems could be a useful tool for studying epigenetic histone modification and chromatin functions.

Apart from histone-mediated epigenetic modification, DNA-binding polyamides can also regulate DNA methylation. Our laboratory reported that a polyamide targeting 5′-CGCG-3′ displayed about 3-fold higher binding affinity for methylated 5′-^{m}CG^{m}CG-3′ than the unmethylated sequence [78a]. Dervan and coworkers reported that minor groove-binding polyamides could inhibit DNA methyltransferase activity in the major groove at their target site [16b]. Kaneda and coworkers successfully inhibited DNA methylation of the *MLH1* promoter in a sequence-specific way

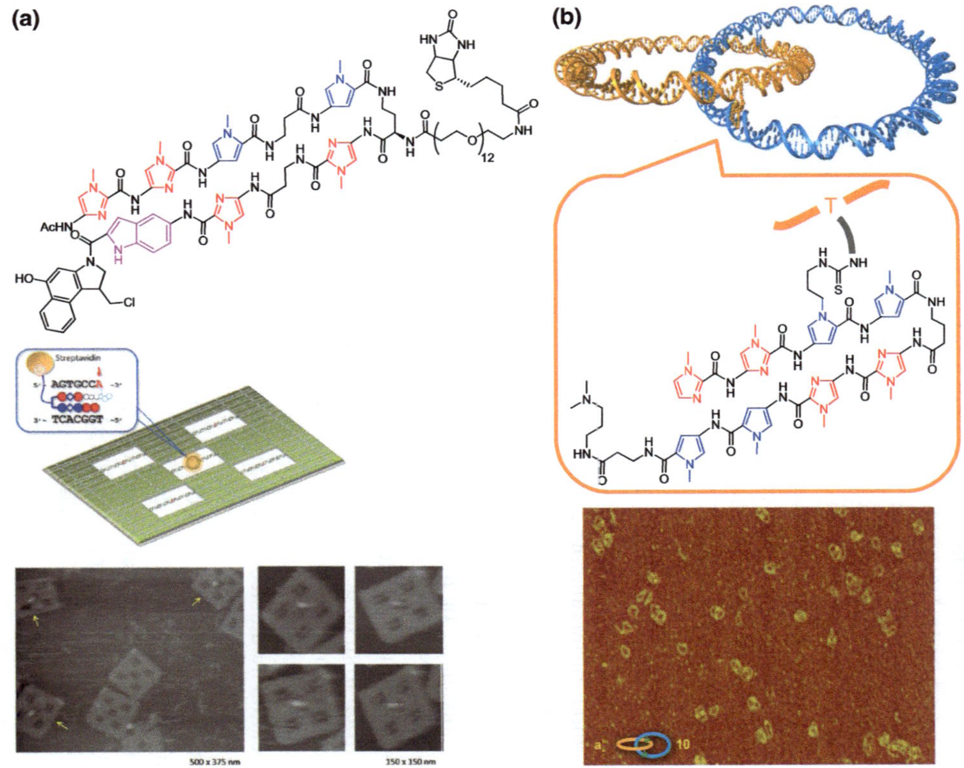

Fig. 1.18 Schematic representation and AFM images of DNA nanostructures regulated by Py–Im polyamides. **a** Imaging of sequence-selective single-molecule alkylation with the biotin-labeled polyamide-*seco*-CBI conjugate and streptavidin. **b** DNA catenane formed by the interaction of one circular DNA–Py–Im polyamide that is covalently attached to thymine of another circular DNA. The images are modified from Ref. [72c, f] respectively. Copyright 2012 American Chemical Society and Copyright 2011 American Chemical Society. Chemical structures are quoted from Ref. [10i]. Copyright 2018 Elsevier

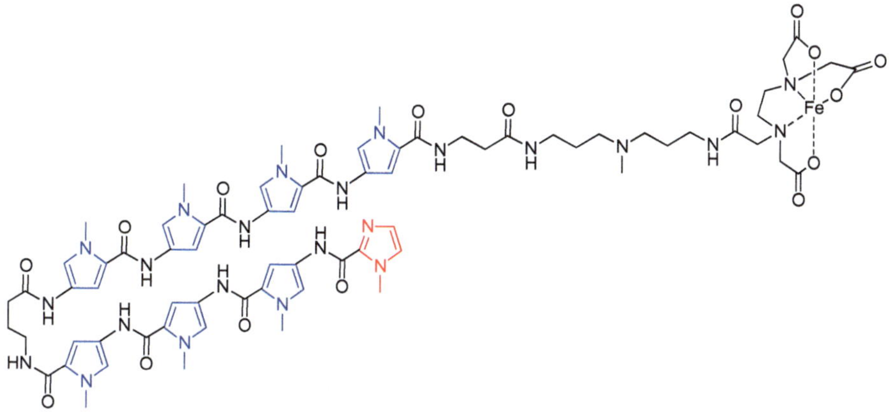

Fig. 1.19 Structure of the Py–Im polyamide–EDTA–Fe(II) conjugate that can cleave DNA in a sequence-specific manner. The image is quoted from Ref. [10i]. Copyright 2018 Elsevier

in RKO human colorectal cancer cells. By means of this methylation inhibition, they succeeded in inhibiting the silencing of *MLH1* transcription [78b]. This result indicates that Py–Im polyamides can act as sequence-specific antagonists of CpG methylation in living cells.

Py–Im polyamides conjugated with a DNA intercalator enforce two different binding modes: minor-groove binding and intercalation [8g, 64, 79]. TO conjugates could mediate sequence-specific enhancement of fluorescence [64], but other applications of DNA intercalator conjugates have been investigated. Polyamide–acridine orange conjugates were reported to unwind DNA and inhibit GCN4 bZIP protein binding by inducing allosteric change in the protein–DNA complex (Fig. 1.21) [79a]. A polyamide–naphthalimide conjugate targeting the sequence flanking the ICB-2 site within the *Topo IIα* promoter was reported to inhibit NF-Y binding and enhance the expression of *Topo IIα* [79b]. These results indicate that Py–Im polyamides conjugated to DNA intercalators also have the potential to change the structure of DNA, inhibit transcription factors, and regulate gene expression.

The sequence specificity of Py–Im polyamides has been evaluated by comparison of the affinity of binding to the matched sequence with that of nonspecific binding, mainly using footprinting [80], PAGE of dsDNA cleaved at alkylation sites [50, 51, 81], thermal denaturation assays of the DNA–polyamide complex [82], surface plasmon analysis (SPR) [5], and the titration of fluorescently labeled polyamide [62]. However, these methods can provide data only on the designated matched and mismatched sequences. Information about sequence specificity within larger sequence contexts is required for further application of Py–Im polyamides. Van Dyke's group developed a selection method using restriction endonuclease protection, selection, and amplification (REPSA), i.e., SELEX. In this method, Py–Im polyamides are incubated with randomized DNA, followed by DNA cleavage with a type IIS restriction enzyme (IISRE), which is inhibited in the presence of bound polyamides. DNA

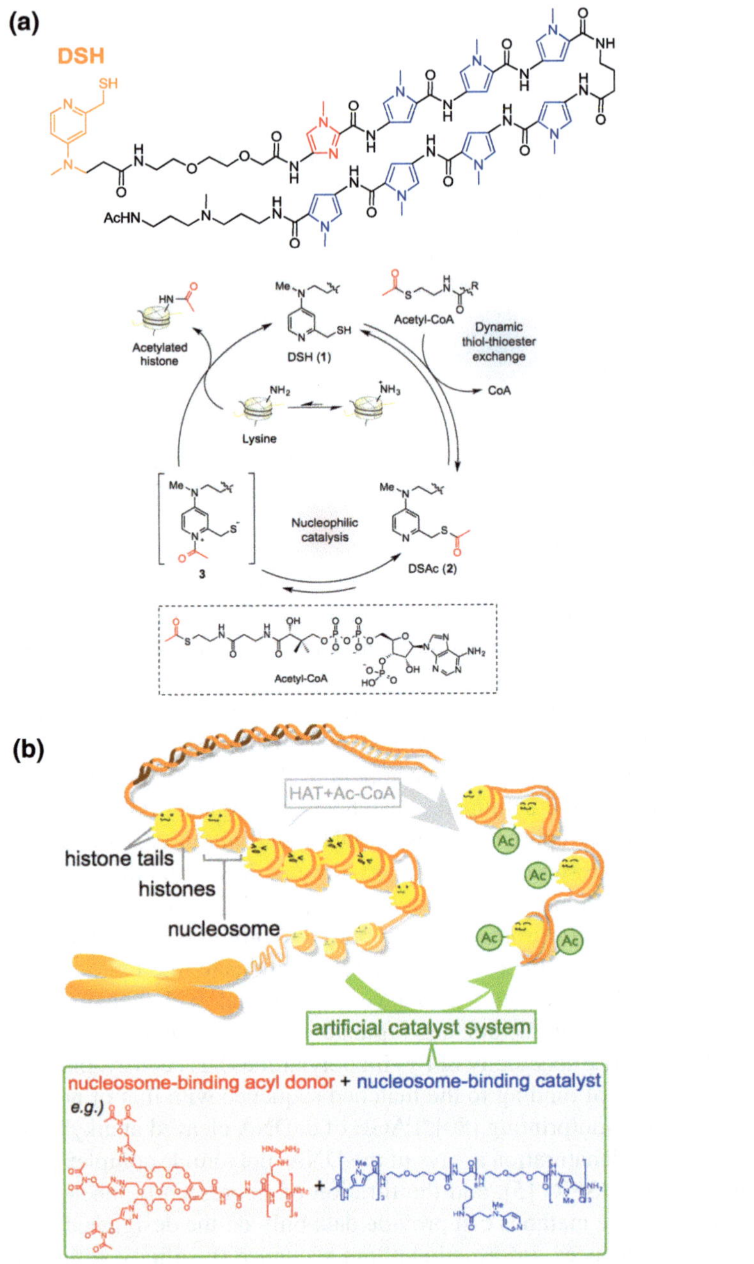

Fig. 1.20 Synthetic nucleosome posttranslational modification system. **a** Structure and reaction mechanism of the Py–Im polyamide–catalytic DSH conjugate. This figure is quoted from Refs. [10i, 77a]. Copyright 2018 Elsevier and Copyright 2017 American Chemical Society. **b** Schematic of histone modification and regulation of chromatin structure by a nucleosome-binding acyl donor and nucleosome-binding catalyst. This figure is reproduced from Ref. [77b]. Copyright 2017 Elsevier

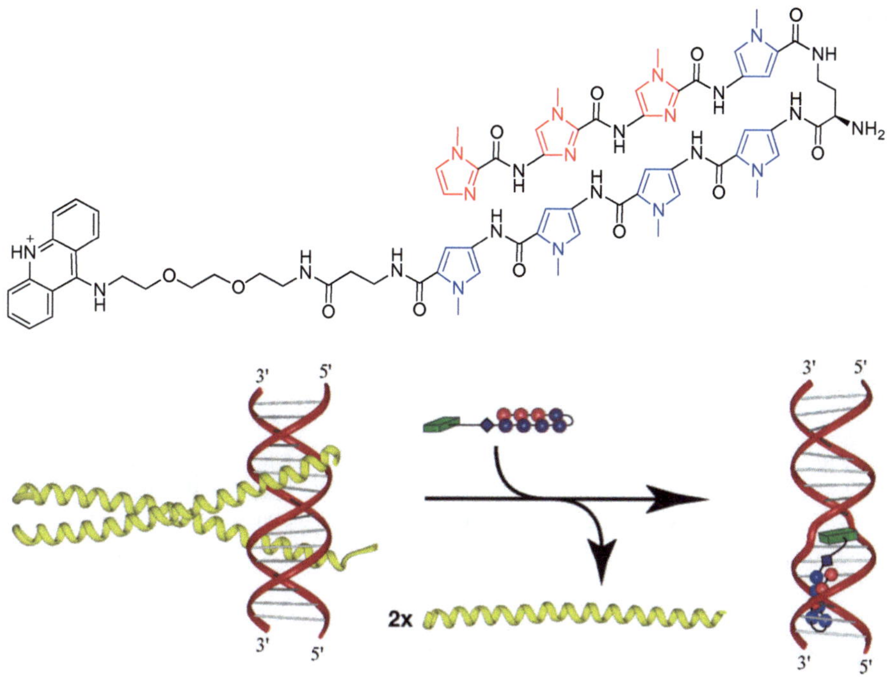

Fig. 1.21 Structure of the Py–Im polyamide–acridine orange intercalator conjugate and the model for allosteric inhibition of a protein–DNA complex by the conjugate. This figure is modified from Ref. [79a]. Copyright 2003 American Chemical Society

that was not cleaved by polyamides could be amplified by PCR [83]. These three steps were repeated to determine the binding sequence of the Py–Im polyamides. However, this REPSA process is low throughput. Ansari, Dervan and coworkers established a cognate site identifier (CSI) microarray platform, in which Cy3-labeled Py–Im polyamides were applied to a microarray to obtain a comprehensive ligand-binding profile (Fig. 1.22a) [84]. However, the need for custom synthesis and analysis of arrays hindered the routine application of this microarray-based method. Using next-generation sequencing technology, Dervan, Segal and colleagues developed the Bind-n-Seq technique [16, 85a]. In this methodology, randomized dsDNA was incubated with biotin-labeled Py–Im polyamides and then purified by biotin-streptavidin interaction, followed by sequencing to identify the binding motifs and evaluate the specificity (Fig. 1.22b). We have also used this method to ensure the specificity of Py–Im polyamides [31b, 50g, 85b–d].

However, this Bind-n-Seq method is limited to providing data for binding under noncellular conditions. Ansari and coworkers established the COSMIC technique using both CSI microarray and next-generation sequencing platforms [86]. As shown in Fig. 1.23a, living cells were treated with Py–Im polyamides conjugated to biotin and the photo-crosslinker psoralen. After using this to crosslink molecules in living

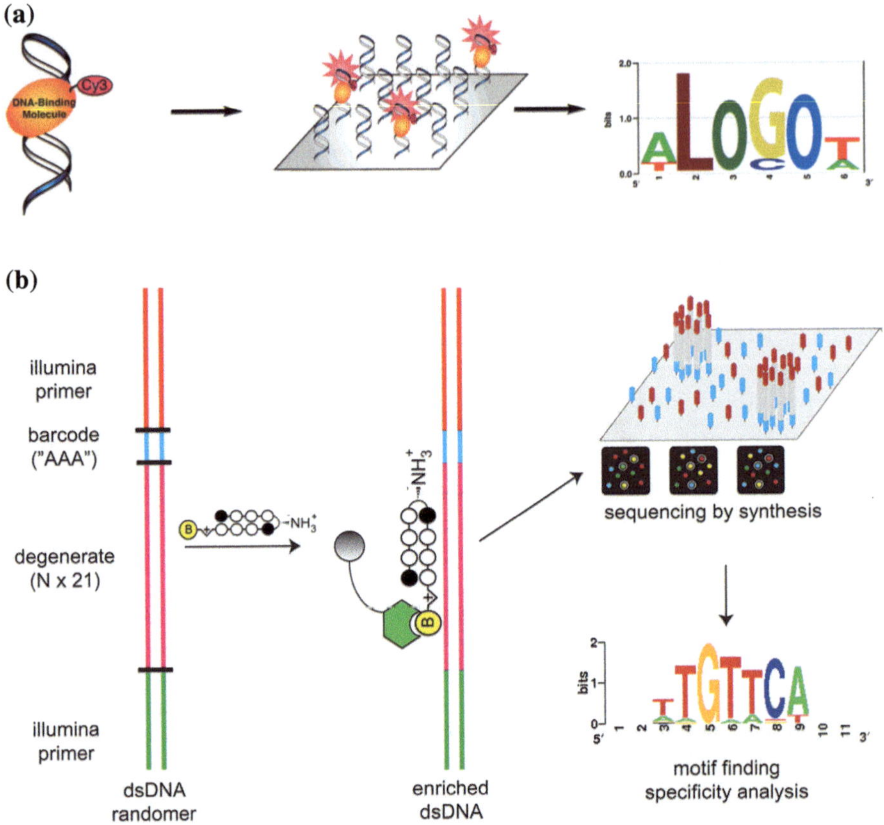

Fig. 1.22 Schematic of the method for evaluation of sequence specificity in the context of longer DNA. **a** CSI microarray platform using Cy3-labeled Py–Im polyamides and DNA microarray. This figure is reproduced from Ref. [84b]. Copyright 2007 American Chemical Society. **b** Bind-n-Seq platform using biotin-labeled Py–Im polyamides and next-generation sequencer. This figure is reproduced from Ref. [16a]. Copyright 2012 American Chemical Society

cells, the cells were lysed and genomic DNA was sheared. Polyamide–DNA adducts were extracted through a biotin–streptavidin interaction, then the DNA bound to the polyamides was released and sequenced. In addition, we recently developed a method to determine the binding sites of Py–Im polyamides in a biologically active, histone-packed, chromatinized genome that was extracted from cells using next-generation sequencers (Fig. 1.23b) [36f, 50h, 87]. The extracted chromatinized genome was incubated with biotin-labeled Py–Im polyamides. Then, the nuclei were treated with micrococcal nuclease to produce mononucleosomes to generate the optimally sized DNA for construction of a sequencing library. Nuclear proteins were digested and then DNA bound to the biotinylated Py–Im polyamides was extracted through a biotin–streptavidin interaction. The extracted and purified DNA was then subjected to high-throughput sequencing. This method could be applied to those polyamides

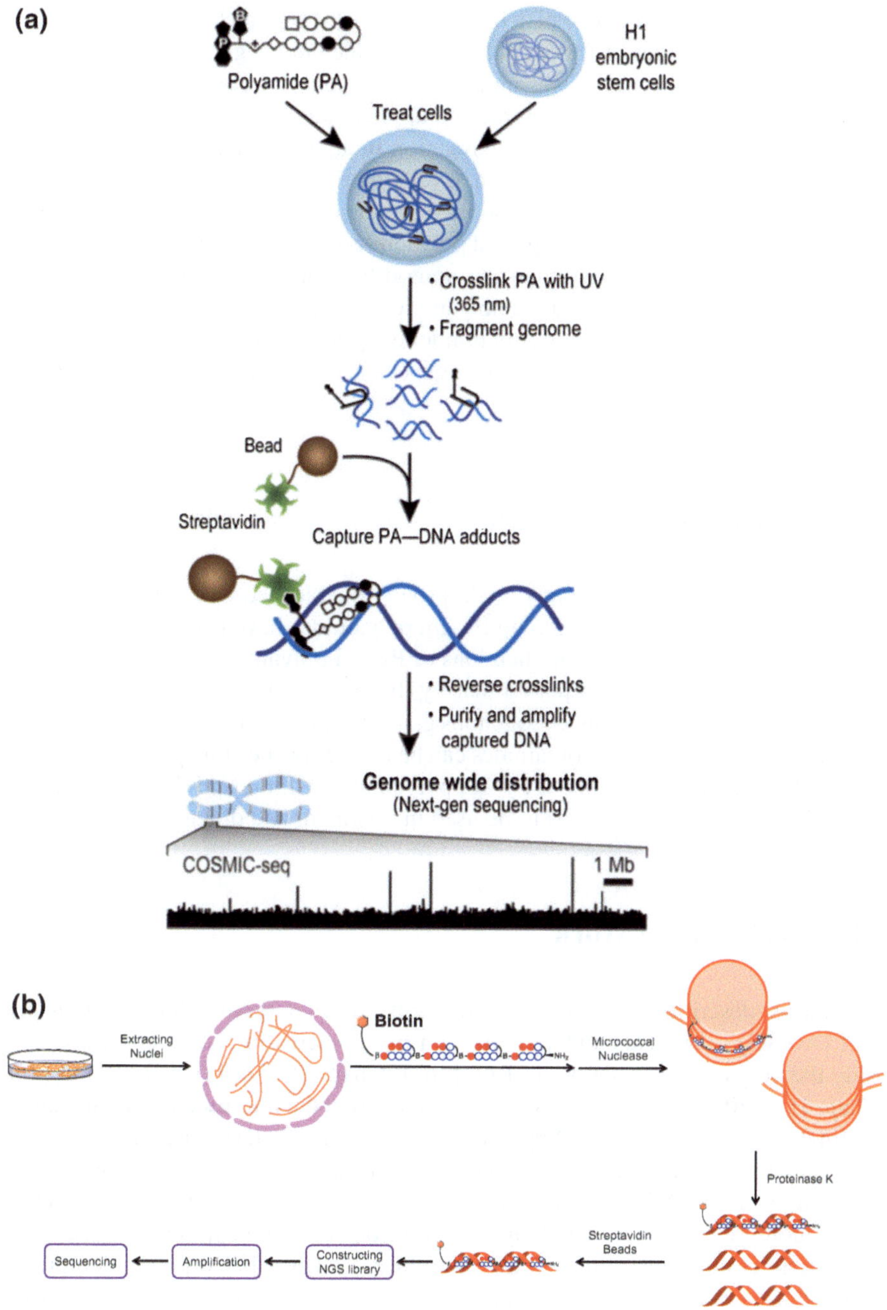

Fig. 1.23 Schematic of the identification of Py–Im polyamide-binding sites in a genomic context. **a** Analysis of DNA crosslinked with Py–Im polyamides in living cells. This figure is modified from Ref. [86b]. Copyright 2016 National Academy of Sciences. **b** Analysis of DNA binding to Py–Im polyamides in the extracted nuclei. This figure is reproduced from Ref. [36f]. Copyright 2016 American Chemical Society

that have difficulty entering nuclei in living cells [36f]. These techniques allowed us to identify the binding sites of Py–Im polyamides in the human genome and to confirm that their sequence specificity was maintained in a genomic context [36f, 86, 87].

Although many interesting features and applications of Py–Im polyamides have been reported, poor solubility and aggregation in aqueous media should be solved for biological and clinical investigation. Cyclodextrin was reported to increase their aqueous solubility and bioefficacy and improve the condition of injection into mice [67e, f]. In addition, our group reported polyamide-PEGylated *seco*-CBI conjugates had higher aqueous solubility than the unmodified ones and showed cytotoxicity [67c]. And liposome to encapsulate polyamide-*seco*-CBI conjugates was suggested to work as a carrier for polyamides in tumor xenograft mouse models [50e]. Further investigation of increasing the solubility is expected to achieve clinical trials of Py–Im polyamides.

1.9 Conclusion and Outlook

Inspired by netropsin and distamycin A, programmable, sequence-specific double-stranded DNA minor groove-binding Py–Im polyamides have been generated. Various structures, functions, and applications of Py–Im polyamides have been studied, and many researchers have reported that Py–Im polyamides are permeable to cells and can up- and downregulate endogenous gene expression in a sequence-dependent manner. In addition, Py–Im polyamides can be used as probes for specific sequences in cells. Further investigations of Py–Im polyamides will expand their applicability and will allow these polyamides to be used in treatment and diagnosis of diseases.

1.10 Scope of the Work

This thesis describes studies of Py–Im polyamides targeting human telomeric 5′-(TTAGGG)$_n$-3′ repeats and visualization of telomeres in chemically-fixed cells using the polyamides as shown in Ref. [36b] (Chap. 2), [36d] (Chap. 3) and [36f] (Chap. 4). In 2001, Laemmli and co-workers reported the fluorescent tandem hairpin Py–Im polyamides targeting 12 bp in telomeric repeats to stain vertebrate telomeres in chemically-fixed cells [36a]. This technique did not require denaturation involved in a widely used PNA-FISH method and could keep the structure of telomeres. Thus, fluorescent signals of polyamides were detected with immunofluorescence of shelterin components. Furthermore, the fluorescently-labeled polyamide probes could be used in estimation of telomere length. Though studies of telomeres using mild staining with the polyamides had been expected, any studies using them had not been reported for a long time because of its synthetic difficulty. Chapter 2 describes the development of their facile synthetic method and successful visualization of human and mouse telomeres under mild conditions and allow easier labeling of telomeres in the cells while maintaining the telomere structure. Furthermore, the telomere length at a single telomere level is found to be related to the abundance of TRF1 protein, a shelterin complex component in the telomere. Chapter 3 describes tandem trimer Py–Im polyamides targeting 18 bp. They are composed of three hairpin moieties and they allow us to visualize human telomeres with higher specificity. Chapter 4 describes further extension of target sites and development of tandem tetramer Py–Im polyamides targeting 24 bp in telomeric repeats, the longest reported binding site for synthetic, non-nucleic-acid-based, sequence-specific DNA-binding molecules. Fluorescently labeled tandem tetramer polyamide probes could visualize human telomeres in chemically fixed cells with higher specificity than tandem dimer and trimer polyamide probes. High-throughput sequencing of human genomic DNA pulled down by the biotin-labeled tandem tetramer polyamide probe confirmed its effective binding to telomeric repeats in the complex chromatinized genome.

References

1. Watson JD, Crick FHC (1953) Nature 171:737–738
2. Rhodes D, Lipps HJ (2015) Nucleic Acids Res 43:8627–8637
3. (a) Finlay AC, Hochstein FA, Sobin BA, Murphy FX (1951) J Am Chem Soc 73:341–343. (b) Arcamone F, Penco P, Orezzi P, Nicolella V, Pirelli A (1964) Nature 203:1064–1065
4. Kopka ML, Yoon C, Goodsell D, Pjura P, Dickerson RE (1985) Proc Natl Acad Sci U S A 82:1376–1380
5. (a) Coll M, Frederick CA, Wang AH-J, Rich A (1987) Proc Natl Acad Sci U S A 84:8385–8389. (b) Pelton JG, Wemmer DE (1989) Proc Natl Acad Sci U S A 86:5723–5727
6. (a) Wade WS, Mrksich M, Dervan PB (1992) J Am Chem Soc 114:8783–8794. (b) Mrksich M, Wade WS, Dwyer TJ, Geierstanger BH, Wemmer DE, Dervan PB (1992) Proc Natl Acad Sci U S A 89:7586–7590. (c) Wade WS, Mrksich M, Dervan PB (1993) Biochemistry 32:11385–11389
7. Mrksich M, Dervan PB (1993) J Am Chem Soc 115:2572–2576
8. (a) Mrksich M, Dervan PB (1994) J Am Chem Soc 115:9892–9899. (b) Mrksich M, Dervan PB (1994) J Am Chem Soc 116:3663–3664. (c) Chen YH, Lown JW (1994) J Am Chem Soc 116:6995–7005. (d) Greenberg WA, Baird EE, Dervan PB (1998) Chem Eur J 4:796–805. (e) O'Hare CC, Mack D, Tandon M, Sharma SK, Lown JW, Kopka ML, Dickerson RE, Hartley JA (2002) Proc Natl Acad Sci U S A 99:72–77. f Olenyuk B, Jitianu C, Dervan PB (2003) J Am Chem Soc 125:4741–4751. (g) Fechter EJ, Olenyuk B, Dervan PB (2004) Angew Chem Int Ed 43:3591–3594. (h) O'Hare CC, Uthe P, Mackay H, Blackmon K, Jones J, Brown T, Nguyen B, Wilson WD, Lee M, Hartley JA (2007) Biochemistry 46:11661–11670. (i) Mackay H, Brown T, Uthe PB, Westrate L, Sielaff A, Jones J, Lajiness JP, Kluza J, O'Hare C, Nguyen B, Davis Z, Bruce C, Wilson WD, Hartley JA, Lee M (2008) Bioorg. Med. Chem. 16:9145–9153. (j) Franks A, Tronrud C, Kiakos K, Kluza J, Munde M, Brown T, Mackay H, Wilson WD, Hochhauser D, Hartley JA, Lee M (2010) Bioorg Med Chem 18:5553–5561. (k) Guo C, Asamitsu S, Kashiwazaki G, Sato S, Bando T, Sugiyama H (2017) ChemBioChem 18:166–170
9. (a) Mrksich M, Parks ME, Dervan PB (1994) J Am Chem Soc 116:7983–7988. (b) Baird EE, Dervan PB (1996) J Am Chem Soc 118:6141–6146. (c) Wurtz NR, Turner JM, Baird EE, Dervan PB (2001) Org Lett 3:1201–1203. (d) Trauger JW, Baird EE, Dervan PB (1996) Nature 382:559–561

10. For reviews: (a) Dervan PB (2001) Bioorg Med Chem 9:2215–2235. (b) Dervan PB, Edelson BS (2003) Curr Opin Struct Biol 13:284–299. (c) Melander C, Burnett R, Gottesfeld JM (2004) J Biotechnol 112:195–220. (d) Murty MSRC, Sugiyama H (2004) Biol Pharm Bull 27:468–474. (e) Dervan PB, Doss RM, Marques MA (2005) Curr Med Chem: Anti-Cancer Agents 5:373–387. (f) Bando T, Sugiyama H (2006) Acc Chem Res 39, 935–944. (g) Blackledge MS, Melander C (2013) Bioorg Med Chem 21:6101–6114. (h) Pandian GN, Sugiyama H (2016) Bull Chem Soc Jpn 89:843–868. (i) Kawamoto Y, Bando T, Sugiyama H (2018) Bioorg Med Chem 26:1393–1411
11. Kielkopf CL, Baird EE, Dervan PD, Rees DC (1998) Nat Struct Biol 5:104–109
12. (a) Chenoweth DM, Dervan PB (2009) Proc Natl Acad Sci U S A 106:13175–13179. (b) Chenoweth DM, Dervan PB (2010) J Am Chem Soc 132:14521–14529
13. (a) Yang X-L, Hubbard RB, Lee M, Tao Z-F, Sugiyama H, Wang AH-J (1999) Nucleic Acids Res 27:4183–4190. (b) Lacy ER, Cox KK, Wilson WD, Lee M (2002) Nucleic Acids Res 30:1834–1841
14. Swalley SE, Baird EE, Dervan PB (1999) J Am Chem Soc 121:1113–1120
15. (a) Herman DM, Baird EE, Dervan PB (1998) J Am Chem Soc 120:1382–1391. (b) Rucker VC, Melander C, Dervan PB (2003) Helv. Chim. Acta 86:1839–1851
16. (a) Meier JL, Yu AS, Korf I, Segal DJ, Dervan PB (2012) J Am Chem Soc 134:17814–17822. (b) Kang JS, Meier JL, Dervan PB (2014) J Am Chem Soc 136:3687–3694
17. Dose C, Ho D, Gaub HE, Dervan PB, Albrecht CH (2007) Angew Chem Int Ed 46:8384–8387
18. Dose C, Farkas ME, Chenoweth DM, Dervan PB (2008) J Am Chem Soc 130:6859–6866
19. Parks ME, Baird EE, Dervan PB (1996) J Am Chem Soc 118:6147–6152
20. (a) Turner JM, Baird EE, Dervan PB (1997) J Am Chem Soc 119:7636–7644. (b) Turner JM, Swalley SE, Baird EE, Dervan PB (1998) J Am Chem Soc 120:6219–6226
21. (a) White S, Szewczyk JW, Turner JM, Baird EE, Dervan PB (1998) Nature 391:468–471. (b) Kielkopf CL, White S, Szewczyk JW, Turner JM, Baird EE, Dervan PB, Rees DC (1998) Science 282:111–115
22. (a) Floreancig PE, Swalley SE, Trauger JW, Dervan PB (2000) J Am Chem Soc 122:6342–6350. (b) Woods CR, Ishii T, Wu B, Bair KW, Boger DL (2002) J Am Chem Soc 124:2148–2152. (c) Zhang W, Bando T, Sugiyama H (2006) J Am Chem Soc 128:8766–8776. (d) Zhang W, Jiang S-K, Wu Y-L, Guo C-X, Zhang H-F, Sugiyama H, Chen X-L (2012) ChemBioChem 13:47–50. (e) Zhang W, Minoshima M, Sugiyama H (2006) J Am Chem Soc 128:14905–14912
23. (a) Zhan Z-YJ, Dervan PB (2000) Bioorg Med Chem 8:2467–2474. (b) Foister S, Marques MA, Doss RM, Dervan PB (2003) Bioorg Med Chem 11:4333–4340. (c) Padroni G, Parkinson JA, Fox KR, Burley GA (2018) Nucleic Acids Res 46:42–53. (d) Anthony NG, Johnston BF, Khalaf AI, MacKay SP, Parkinson JA, Suckling CJ, Waigh RD (2004) J Am Chem Soc 126:11338–11349. (e) James PL, Merkina EE, Khalaf AI, Suckling CJ, Waigh RD, Brown T, Fox KR (2004) Nucleic Acids Res 32:3410–3417. (f) Nguyen DH, Szewczyk JW, Baird EE, Dervan PB (2001) Bioorg Med Chem 9:7–17. (g) Marques MA, Doss RM, Urbach AR, Dervan PB (2002) Helv Chim Acta 85:4485–4517. (h) Muzikar KA, Meier JL, Gubler DA, Raskatov JA, Dervan PB (2011) Org Lett 13:5612–5615
24. (a) Renneberg D, Dervan PB (2003) J Am Chem Soc 125:5707–5716. (b) Briehn CA, Weyermann P, Dervan PB (2003) Chem Eur J 9:2110–2122. (c) Marques MA, Doss RM, Foister S, Dervan PB (2004) J Am Chem Soc 126:10339–10349. (d) Doss R, Marques MA, Foister S, Chenoweth DM, Dervan PB (2006) J Am Chem Soc 128:9074–9079. (e) Chenoweth DM, Viger A, Dervan PB (2007) J Am Chem Soc 129:2216–2217
25. (a) Lacy ER, Le NM, Price CA, Lee M, Wilson WD (2002) J Am Chem Soc 124:2153–2163. (b) Buchmueller KL, Staples AM, Uthe PB, Howard CM, Pacheco KA, Cox KK, Henry JA, Bailey SL, Horick SM, Nguyen B, Wilson WD, Lee M (2005) Nucleic Acids Res 33:912–921. (c) Chavda S, Liu Y, Babu B, Davis R, Sielaff A, Ruprich J, Westrate L, Tronrud C, Ferguson A, Franks A, Tzou S, Adkins C, Rice T, Mackay H, Kluza J, Tahir SA, Lin S, Kiakos K, Bruce CD, Wilson WD, Hartley JA, Lee M (2011) Biochemistry 50:3127–3136. (d) Satam V, Babu B, Patil P, Brien KA, Olson K, Savagian M, Lee M, Mepham A, Jobe LB, Bingham JP, Pett L, Wang S, Ferrara M, Bruce CD, Wilson WD, Lee M, Hartley JA, Kiakos K (2015) Bioorg Med Chem Lett 25:3681–3685

26. Hawkins CA, Pela´ez de Clairac R, Dominey RN, Baird EE, White S, Dervan PB, Wemmer DE (2000) J Am Chem Soc 122:5235–5243
27. (a) Trauger JW, Baird EE, Mrksich M, Dervan PB (1996) J Am Chem Soc 118:6160–6166. (b) Swalley SE, Baird EE, Dervan PB (1997) Chem Eur J 3:1600–1607. (c) Trauger JW, Baird EE, Dervan PB (1998) J Am Chem Soc 120:3534–3535
28. (a) Janssen S, Durussel T, Laemmli UK (2000) Mol Cell 6:999–1011. (b) Janssen S, Cuvier O, Müller M, Laemmli UK (2000) Mol Cell 6:1013–1024. (c) Blattes R, Monod C, Susbielle G, Cuvier O, Wu J-H, Hsieh T-S, Laemmli UK, Käs E (2006) EMBO J 25:2397–2408. (d) Urbach AM, Dervan PB (2001) Proc Natl Acad Sci U S A 98:4343–4348. (e) Urbach AM, Love JJ, Ross SA, Dervan PB (2002) J Mol Biol 320:55–71. (f) Burnett R, Melander C, Puckett JW, Son LS, Wells RD, Dervan PB, Gottesfeld JM (2006) Proc Natl Acad Sci U S A 103:11497–11502. (g) Wang RE, Pandita RK, Cai J, Hunt CR, Taylor J-S (2012) ChemBioChem 13:97–104. (h) Du J, Campau E, Soragni E, Ku S, Puckett JW, Dervan PB, Gottesfeld JM (2012) J Biol Chem 287:29861–29872. (i) Gerhardt J, Bhalla AD, Butler JS, Puckett JW, Dervan PB, Rosenwaks Z, Napierala, M. Cell Rep 16:1218–1227
29. Melander C, Herman DM, Dervan PB (2000) Chem Eur J 6:4487–4497
30. (a) Cho JY, Parks MP, Dervan PB (1995) Proc Natl Acad Sci U S A 92:10389–10392. (b) Herman DM, Turner JM, Baird EE, Dervan PB (1999) J Am Chem Soc 121:1121–1129. (c) Melander C, Herman DM, Dervan PB (2000) Chem Eur J 6:4487–4497. (d) Zhang Q, Dwyer TJ, Tsui V, Case DA, Cho J, Dervan PB, Wemmer DE (2004) J Am Chem Soc 126:7958–7966. (e) Chenoweth DM, Harki DA, Phillips JW, Dose C, Dervan PB (2009) J Am Chem Soc 131:7182–7188. (f) Li BC, Montgomery DC, Puckett JW, Dervan PB (2013) J Org Chem 78:124–133
31. (a) Morinaga H, Bando T, Takagaki T, Yamamoto H, Hashiya K, Sugiyama H (2011) J Am Chem Soc 133:18924–18930. (b) Kashiwazaki G, Chandran A, Asamitsu S, Kawase T, Kawamoto Y, Sawatani Y, Hashiya K, Bando T, Sugiyama H (2016) ChemBioChem 17:1752–1758
32. Yamamoto M, Bando T, Morinaga N, Kawamoto Y, Hashiya K, Sugiyama H (2014) Chem Eur J 20:752–759
33. Weyermann P, Dervan PB (2002) J Am Chem Soc 124:6872–6878
34. (a) Edayathumangalam RS, Weyermann P, Gottesfeld J. M, Dervan, P. B, Luger, K. Proc Natl Acad Sci U S A 101:6864–6869. (b) Edayathumangalam RS, Weyermann P, Dervan, PB, Gottesfeld JM, Luger K (2005) J Mol Biol 345:103–114. (c) Obata S, Asamitsu S, Hashiya K, Bando T, Sugiyama H (2018) Biochemistry 57:498–502
35. (a) Herman DM, Baird EE, Dervan PB (1999) Chem Eur J 5:975–983. (b) Kers I, Dervan PB (2002) Bioorg Med Chem 10:3339–3349. (c) Schaal TD, Mallet WG, McMinn DL, Nguyen NV, Sopko MM, John S, Parekh BS (2003) Nucleic Acids Res 31:1282–1291. (d) Sasaki S, Bando T, Minoshima M, Shinohara K, Sugiyama H (2008) Chem Eur J 14:864–870. (e) Yamamoto M, Bando T, Kawamoto Y, Taylor RD, Hashiya K, Sugiyama H (2014) Bioconjugate Chem 25:552–559. (f) Taylor RD, Kawamoto Y, Hashiya K, Bando T, Sugiyama H (2014) Chem. Asian J 9:2527–2533
36. (a) Maeshima K, Janssen S, Laemmli UK (2001) EMBO J 20:3218–3228. (b) Kawamoto Y, Bando T, Kamada F, Li Y, Hashiya K, Maeshima K, Sugiyama H (2013) J Am Chem Soc 135:16468–16477. (c) Hirata A, Nokihara K, Kawamoto Y, Bando T, Sasaki A, Ide S, Maeshima K, Kasama T, Sugiyama H (2014) J Am Chem Soc 136:11546–11554. (d) Kawamoto Y, Sasaki A, Hashiya K, Ide S, Bando T, Maeshima K, Sugiyama H (2015) Chem. Sci. 6:2307–2312. (e) Sasaki A, Ide S, Kawamoto Y, Bando T, Murata Y, Shimura M, Yamada K, Hirata A, Nokihara K, Hirata T, Sugiyama H, Maeshima K (2016) Sci Rep 6:29261. (f) Kawamoto Y, Sasaki A, Chandran A, Hashiya K, Ide S, Bando T, Maeshima K, Sugiyama H (2016) J Am Chem Soc 138:14100–14107
37. (a) Best TP, Edelson BS, Nickols NG, Dervan PB (2003) Proc Natl Acad Sci U S A 100:12063–12068. (b) Edelson BS, Best TP, Olenyuk B, Nickols NG, Doss RM, Foister S, Heckel A, Dervan PB (2004) Nucleic Acids Res 32:2802–2818
38. Gottesfeld JM, Neely L, Trauger JW, Baird EE, Dervan PB (1997) Nature 391:468–471

39. Dickinson LA, Gulizia RJ, Trauger JW, Baird EE, Mosier DE, Gottesfeld JM, Dervan PB (1998) Proc Natl Acad Sci U S A 95:12890–12895
40. (a) Olenyuk BZ, Zhang GJ, Klco JM, Nickols NG, Kaelin WG Jr, Dervan PB (2004) Proc Natl Acad Sci U S A 101:16768–16773. (b) Nickols NG, Jacobs CS, Farkas ME, Dervan PB (2007) Nucleic Acids Res 35:363–370. (c) Nickols NG, Jacobs CS, Farkas ME, Dervan PB (2007) ACS Chem Biol 2:561–571. (d) Szablowski JO, Raskatov JA, Dervan PB (2016) Mol Cancer Ther 15:608–617. (e) Mysore VS, Szablowski JO, Dervan PB, Frost PJ (2016) Mol Cancer Res 14:253–266
41. (a) Nickols NG, Dervan PB (2007) Proc Natl Acad Sci U S A 104:10418–10423. (b) Hargrove AE, Martinez TF, Hare AA, Kurmis AA, Phillips JW, Sud S, Pienta KJ, Dervan PB (2015) PLoS ONE 10:143161. (c) Yang F, Nickols NG, Li BC, Marinov G, Said JW, Dervan PB (2013) Proc Natl Acad Sci U S A 110:1863–1868. (d) Muzikar KA, Nickols NG, Dervan PB (2009) Proc Natl Acad Sci U S A 106:16598–16603 (e) Kurmis AA, Yang F, Welch TR, Nickols NG, Dervan PB (2017) Cancer Res 77:2207–2212. (f) Martínez TF, Phillips JW, Karanja KK, Polaczek P, Wang CM, Li BC, Campbell JL, Dervan PB (2014) Nucleic Acids Res 42:11546–11559
42. (a) Lai Y-M, Fukuda N, Ueno T, Kishioka H, Matsuda H, Saito S, Matsumoto K, Ayame H, Bando T, Sugiyama H, Mugishima H, Serie K (2005) J Pharmacol Exp Ther 315:571–575. (b) Matsuda H, Fukuda N, Ueno T, Tahira Y, Ayame H, Zhang W, Bando T, Sugiyama H, Saito S, Matsumoto K, Mugishima H, Serie K (2006) J Am Soc Nephrol 17:422–432. (c) MatsudaH, Fukuda N, Ueno T, Katakawa M, Wang X, Watanabe T, Matsui S, Aoyama T, Saito K, Bando T, Matsumoto Y, Nagase H, Matsumoto K, Sugiyama H (2011) Kidney Int 79:46–56. (d) Yao E-H, Fukuda N, Ueno T, Matsuda H, Nagase H, Matsumoto Y, Sugiyama H, Matsumoto K (2009) Cardiovasc Res 81:797–804. (e) Chen M, Matsuda H, Wang L, Watanabe T, Kimura MT, Igarashi J, Wang X, Sakimoto T, Fukuda N, Sawa M, Nagase H (2010) Mol Ther 18:519–527. (f) Washio H, Fukuda N, Matsuda H, Nagase H, Watanabe T, Matsumoto Y, Terui T (2011) J Invest Dermatol 131:1987–1995. (g) Igarashi J, Fukuda N, Inoue T, Nakai S, Saito K, Fujiwara K, Matsuda H, Ueno T, Matsumoto Y, Watanabe T, Nagase H, Bando T, Sugiyama H, Itoh T, Soma M (2015) PLoS ONE 10:125295
43. (a) Hochhauser D, Kotecha M, O'hare C, Morris PJ, Hartley JA, Taherbhai Z, Harris D, Forni C, Mantovani R, Lee M, Hartley JA (2007) Mol Cancer Ther 6:346–354. (b) Kiakos K, Pett L, Satam V, Patil P, Hochhauser D, Lee M, Hartley JA (2015) Chem Biol 22:862–875
44. Taniguchi J, Pandian GN, Hidaka T, Hashiya K, Bando T, Kim KK, Sugiyama H (2017) Nucleic Acids Res 45:9219–9228
45. Hidaka T, Pandian G, Taniguchi J, Nobeyama T, Hashiya K, Bando T, Sugiyama H (2017) J Am Chem Soc 139:8444–8447
46. (a) Edwards TG, Koeller KJ, Slomczynska U, Fok K, Helmus M, Bashkin JK, Fisher C (2011) Antivir Res 91:177–186. (b) Castaneda CH, Scuderi MJ, Edwards TG, Harris Jr GD, Dupureur CM, Koeller KJ, Fisher C, Bashkin JK (2016) Med Chem Commun 7:2076–2082
47. (a) Oyoshi T, Kawakami W, Narita A, Bando T, Sugiyama HJ (2003) Am Chem Soc 125:4752–4754. (b) Asamitsu S, Kawamoto Y, Hashiya F, Hashiya K, Yamamoto M, Kizaki S, Bando T, Sugiyama H (2014) Bioorg Med Chem 22:4646–4657
48. (a) Wurtz NR, Dervan PB (2000) Chem Biol 7:153–161. (b) Wang YD, Dziegielewski J, Wurtz NR, Dziegielewska B, Dervan PB, Beerman TA (2003) Nucleic Acids Res 31:1208–1215. (c) Dickinson LA, Burnett R, Melander C, Edelson BS, Arora PS, Dervan PB, Gottesfeld JM (2004) Chem Biol 11:1583–1594. (d) Alvarez D, Chou CJ, Latella L, Zeitlin SG, Ku S, Puri PL, Dervan PB, Gottesfeld JM (2006) Cell Cycle 5:1537–1548. (e) Tsai SM, Farkas ME, Chou CJ, Gottesfeld JM, Dervan PB (2007) Nucleic Acids Res 35:307–316. (f) Chou CJ, Farkas ME, Tsai SM, Alvarez D, Dervan PB, Gottesfeld JM (2008) Mol Cancer Ther 7:769–778. (g) Chou CJ, O'Hare T, Lefebvre S, Alvarez D, Tyner JW, Eide CA, Druker BJ, Gottesfeld JM (2008) PLoS ONE 3:3593. (h) Jespersen C, Soragni E, James Chou C, Arora PS, Dervan PB, Gottesfeld JM (2012) Bioorg Med Chem Lett 22:4068–4071. (i) Morita K, Suzuki K, Maeda S, Matsuo A, Mitsuda Y, Tokushige C, Kashiwazaki G, Taniguchi J, Maeda R, Noura M, Hirata M, Kataoka T, Yano A, Yamada Y, Kiyose H, Tokumasu M, Matsuo H, Tanaka S, Okuno Y,

Muto M, Naka K, Ito K, Kitamura T, Kaneda Y, Liu PP, Bando T, Adachi S, Sugiyama H, Kamikubo Y (2017) J Clin Invest 127:2815–2828
49. (a) Chang AY, Dervan PB (2000) J Am Chem Soc 122:4856–4864. (b) Wang Y-D, Dziegielewski J, Chang AY, Dervan PB, Beerman TA (2002) J Biol Chem 277:42431–42437. (c) Philips BJ, Chang AY, Dervan PB, Beerman TA (2005) Mol Pharmacol 67:877–882
50. (a) Bando T, Sasaki S, Minoshima M, Dohno C, Shinohara K, Narita A, Sugiyama H (2006) Bioconjugate Chem 17:715–720. (b) Shinohara K, Sasaki S, Minoshima M, Bando T, Sugiyama H (2006) Nucleic Acids Res 34:1189–1195. (c) Shinohara K, Bando T, Sasaki S, Sakakibara Y, Minoshima M, Sugiyama H (2006) Cancer Sci 97:219–225. (d) Sasaki S, Bando T, Minoshima M, Shimizu T, Shinohara K, Takaoka T, Sugiyama H (2006) J Am Chem Soc 128:12162–12168. (e) Kashiwazaki G, Bando T, Yoshidome T, Masui S, Takagaki T, Hashiya K, Pandian GN, Yasuoka J, Akiyoshi K, Sugiyama H (2012) J Med Chem 55:2057–2066. (f) Hiraoka K, Inoue T, Taylor RD, Watanabe T, Koshikawa N, Yoda H, Shinohara K, Takatori A, Sugimoto H, Maru Y, Denda T, Fujiwara K, Balmain A, Ozaki T, Bando T, Sugiyama H, Nagase H (2015) Nat Commun 6:6706. (g) Taylor RD, Chandran A, Kashiwazaki G, Hashiya K, Bando T, Nagase H, Sugiyama H (2015) Chem Eur J 21:14996–15003. (h) Lin J, Hiraoka K, Watanabe T, Kuo T, Shinozuka Y, Takatori A, Koshikawa N, Chandran A, Otsuki J, Sugiyama H, Horton P, Nagase H (2016) PLoS ONE 11:165581
51. (a) Tao Z-F, Saito I, Sugiyama H (2000) J Am Chem Soc 122:1602–1608. (b) Bando T, Narita A, Saito I, Sugiyama H (2002) Chem Eur J 8:4781–4790. (c) Bando T, Narita A, Iwai A, Kihara K, Sugiyama H (2004) J Am Chem Soc 126:3406–3407. (d) Shinohara K, Narita A, Oyoshi T, Bando T, Teraoka H, Sugiyama H (2004) J Am Chem Soc 126:5113–5118
52. (a) Mapp AK, Ansari AZ, Ptashne M, Dervan PB (2000) Proc Natl Acad Sci U S A 97:3930–3935. (b) Ansari AZ, Mapp AK, Nguyen DH, Dervan PB, Ptashne M (2001) Chem Biol 8:583–592. (c) Arora PS, Ansari AZ, Best TP, Ptashne M, Dervan PB (2002) J Am Chem Soc 124:13067–13071. (d) Kwon Y, Arndt HD, Mao Q, Choi Y, Kawazoe Y, Dervan PB, Uesugi M (2004) J Am Chem Soc 126:15940–15941. (e) Arndt H-D, Hauschild KE, Sullivan DP, Lake K, Dervan PB, Ansari AZ (2003) J Am Chem Soc 125:13322–13323. (f) Hauschild KE, Metzler RE, Arndt H-D, Moretti R, Raffaelle R, Dervan PB, Ansari AZ (2005) Proc Natl Acad Sci U S A 102, 5008–5013. (g) Stafford RL, Arndt H-D, Brezinski ML, Ansari AZ, Dervan PB (2007) J Am Chem Soc 129:2660–2668. (h) Stafford RL, Dervan PB (2007) J Am Chem Soc 129:14026–14033. (i) Moretti R, Donato LJ, Brezinski ML, Stafford RL, Hoff H, Thorson JS, Dervan PB, Ansari AZ (2008) ACS Chem Biol 3:220–229
53. (a) Xiao XS, Yu P, Lim HS, Sikder D, Kodadek TA (2007) Angew Chem Int Ed 46:2865–2868. (b) Xiao XS, Yu P, Lim HS, Sikder D, Kodadek TA (2007) J Comb Chem 9:592–600
54. (a) Ohtsuki A, Kimura MT, Minoshima M, Suzuki T, Ikeda M, Bando T, Nagase H, Shinohara KI, Sugiyama H (2009) Tetrahedron Lett 50:7288–7292. (b) Pandian GN, Shinohara K, Ohtsuki A, Nakano Y, Minoshima M, Bando T, Nagase H, Yamada Y, Watanabe A, Terada N, Sato S, Morinaga H, Sugiyama H (2011) ChemBioChem 12:2822–2828. (c) Pandian GN, Ohtsuki A, Bando T, Sato S, Hashiya K, Sugiyama H (2012) Bioorg Med Chem 20:2656–2660. (d) Pandian GN, Nakano Y, Sato S, Morinaga H, Bando T, Nagase H, Sugiyama H (2012) Sci Rep 2:544. (e) Pandian GN, Taniguchi J, Junetha S, Sato S, Han L, Saha A, Anandhkumar C, Bando T, Nagase H, Vaijayanthi T, Taylor RD, Sugiyama H (2014) Sci Rep 4:3843. (f) Han L, Pandian GN, Junetha S, Sato S, Anandahkumar C, Taniguchi J, Saha A, Bando T, Nagase H, Sugiyama H (2013) Angew Chem Int Ed 52:13410–13413. (g) Pandian GN, Sato S, Anandhakumar C, Taniguchi J, Takashima K, Syed J, Han L, Saha A, Bando T, Nagase H, Sugiyama H (2014) ACS Chem Biol 9:2729–2736. (h) Syed J, Chandran A, Pandian GN, Taniguchi J, Sato S, Hashiya K, Kashiwazaki G, Bando T, Sugiyama H (2015) ChemBioChem 16:1497–1501. (i) Wei Y, Pandian GN, Zou T, Taniguchi J, Sato S, Kashiwazaki G, Vaijayanthi T, Hidaka T, Bando T, Sugiyama H (2016) ChemistryOpen 5:517–521
55. Han L, Pandian GN, Chandran A, Sato S, Taniguchi J, Kashiwazaki G, Sawatani Y, Hashiya K, Bando T, Xu Y, Qian X, Sugiyama H (2015) Angew Chem Int Ed 54:8700–8703
56. Yu Z, Taniguchi J, Wei Y, Pandian G, Hashiya K, Bando T, Sugiyama H (2017) Eur J Med Chem 138:320–327

57. Erwin GS, Grieshop MP, Ali A, Qi J, Lawlor M, Kumar D, Ahmad I, McNally A, Teider N, Worringer K, Sivasankaran R, Syed DN, Eguchi A, Ashraf M, Jeffery J, Xu M, Park PMC, Mukhtar H, Srivastava AK, Faruq M, Bradner JE, Ansari AZ (2017) Science 358:1617–1622
58. (a) Patel S, Jung D, Yin PT, Carlton P, Yamamoto M, Bando T, Sugiyama H, Lee K-B (2014) ACS Nano 8:8959–8967. (b) Patel S, Yin PT, Sugiyama H, Lee K-B (2015) ACS Nano 9:6909–6917. (c) Patel S, Pongkulapa T, Yin PT, Pandian GN, Rathnam C, Bando T, Vaijayanthi T, Sugiyama H, Lee K-B (2015) J Am Chem Soc 137:4598–4601. (d) Patel S, Chueng S-TD, Yin PT, Dardir K, Song Z, Pasquale N, Kwan K, Sugiyama H, Lee K-B (2015) Angew Chem Int Ed 54:11983–11988
59. For review: Vaijayanthi T, Bando T, Pandian GN, Sugiyama H (2012) ChemBioChem 13:2170–2185
60. Gygi MP, Ferguson MD, Mefford HC, Lund KP, O'Day C, Zhou P, Friedman C, van den Engh G, Stolowitz ML, Trask BJ (2002) Nucleic Acids Res 30:2790–2799
61. Nozeret K, Loll F, Escudé C, Boutorine AS (2015) ChemBioChem 16:549–554
62. (a) Rucker VC, Foister S, Melander C, Dervan PB (2003) J Am Chem Soc 125:1195–1202. (b) Rucker VC, Dunn AR, Sharma S, Dervan PB, Gray HB (2004) J Phys Chem B 108:7490–7494
63. Fujimoto J, Bando T, Minoshima M, Kashiwazaki G, Nishijima S, Shinohara K, Sugiyama H (2008) Bioorg Med Chem 16:9741–9744
64. Fechter EJ, Olenyuk B, Dervan PB (2005) J Am Chem Soc 127:16685–16691
65. (a) Bando T, Fujimoto J, Minoshima M, Shinohara K, Sasaki S, Kashiwazaki G, Mizumura M, Sugiyama H (2007) Bioorg Med Chem 15:6937–6942. (b) Fujimoto J, Bando T, Minoshima M, Uchida S, Iwasaki M, Shinohara K, Sugiyama H (2008) Bioorg Med Chem 16:5899–5907
66. (a) Belitsky JM, Leslie SJ, Arora PS, Beerman TA, Dervan PB (2002) Bioorg Med Chem 10:3313–3318. (b) Nishijima S, Shinohara K, Bando T, Minoshima M, Kashiwazaki G, Sugiyama H (2010) Bioorg Med Chem 18:978–983
67. (a) Jacobs CS, Dervan PB (2009) J Med Chem 52:7380–7388. (b) Liu B, Kodadek T (2009) J Med Chem 52:4604–4612. (c) Takagaki T, Bando T, Kitano M, Hashiya K, Kashiwazaki G, Sugiyama H (2011) Bioorg Med Chem 19:5896–5902. (d) Meier JL, Montgomery DC, Dervan PB (2012) Nucleic Acids Res 40:2345–2356. (e) Hargrove AE, Raskatov JA, Meier JL, Montgomery DC, Dervan PB (2012) J Med Chem 55:5425–5432. (f) Saha A, Pandian GN, Sato S, Taniguchi J, Kawamoto Y, Hashiya K, Bando T, Sugiyama H (2014) ChemMedChem 9:2374–2380
68. (a) Harki DA, Satyamurthy N, Stout DB, Phelps ME, Dervan PB (2008) Proc Natl Acad Sci U S A 105:13039–13044. (b) Raskatov JA, Puckett JW, Dervan PB (2014) Bioorg Med Chem 22:4371–4375. (c) Raskatov JA, Szablowski JO, Dervan PB (2014) J Med Chem 57:8471–8476
69. (a) Su W, Schuster M, Bagshaw CR, Rant U, Burley GA (2011) Angew Chem Int Ed 50:2712–2715. (b) Su W, Bagshaw CR, Burley GA (2013) Sci Rep 3:1883
70. Han Y-W, Tsunaka Y, Yokota H, Matsumoto T, Kashiwazaki G, Morinaga H, Hashiya K, Bando T, Sugiyama H, Harada Y (2014) Biomater. Sci. 2:297–307
71. (a) Morinaga H, Takenaka T, Hashiya F, Kizaki S, Hashiya K, Bando T, Sugiyama H (2013) Nucleic Acids Res 41:4724–4728. (b) Saha A, Hashiya F, Kizaki S, Asamitsu S, Hashiya K, Bando T, Sugiyama H (2015) Chem Commun 51:14485–14488
72. (a) Cohen JD, Sadowski JP, Dervan PB (2007) Angew Chem Int Ed 46:7956–7959. (b) Cohen JD, Sadowski JP, Dervan PBJ (2008) Am Chem Soc 130:402–403. (c) Yoshidome T, Endo M, Kashiwazaki G, Hidaka K, Bando T, Sugiyama H (2012) J Am Chem Soc 134:4654–4660. (d) Schmidt TL, Nandi CK, Rasched G, Parui PP, Brutschy B, Famulok M, Heckel A (2007) Angew Chem Int Ed 46:4382–4384. (e) Schmidt TL, Heckel A (2009) Small 5:1517–1520. (f) Schmidt TL, Heckel A (2011) Nano Lett 11:1739–1742
73. (a) Schultz PG, Taylor JS, Dervan PB (1982) J Am Chem Soc 104:6861–6863. (b) Schultz PG, Dervan PB (1983) Proc Natl Acad Sci U S A 80:6834–6837. (c) Schultz PG, Dervan PB (1983) J Am Chem Soc 105:7748–7750. (d) Taylor JS, Schultz PG, Dervan PB (1984) Tetrahedron 40:457–465. (e) Youngquist RS, Dervan PB (1985) Proc Natl Acad Sci U S A 82:2565–2569. (f) Youngquist RS, Dervan PB (1985) J Am Chem Soc 107:5528–5529. (g) Dervan PB (1986) Science 232:464–471. (h) Youngquist RS, Dervan PB (1987) J Am Chem Soc 109:7564–7566

74. (a) Pitié M, Van Horn JD, Brion D, Burrows CJ, Meunier B (2000) Bioconjugate Chem 11:892–900. (b) Pitié M, Burrows CJ, Meunier B (2000) Nucleic Acids Res 28:4856–4864. (c) Liu D, Zhou J, Li H, Zheng B, Yuan G (2006) Bioorg Med Chem Lett 16:5032–5035. (d) Simon P, Cannata F, Perrouault L, Halby L, Concordet J-P, Boutorine A, Ryabinin V, Sinyakov A, Giovannangeli C (2008) Nucleic Acids Res 36:3531–3538. (e) Li C, Qiao R-Z, Wang Y-Q, Zhao Y-F, Zeng R (2008) Bioorg Med Chem Lett 18:5766–5770. (f) Li C, Du C, Tian H, Jiang C, Du M, Liu Y, Qiao R-Z, Jia Y-X, Zhao Y-F (2010) Chem Eur J 16:12935–12940. (g) Kameshima W, Ishizuka T, Minoshima M, Yamamoto M, Sugiyama H, Xu Y, Komiyama M (2013) Angew Chem Int Ed 52:13681–13684
75. (a) Tanaka H, Tanaka Y, Minoshima M, Yamaguchi S, Fuse S, Doi T, Kawauchi S, Sugiyama H, Takahashi T (2010) Chem Commun 46:5942–5944. (b) Wirth-Hamdoune D, Ullrich S, Scheffer U, Radanović T, Dürner G, Göbel MW (2016) ChemBioChem 17:506–514
76. (a) Wang CCC, Dervan PB (2001) J Am Chem Soc 123:8657–8661. (b) Arimondo PB, Bailly C, Boutorine AS, Ryabinin VA, Syniakov AN, Sun JS, Garestier T, Hélène C (2001) Angew Chem Int Ed 40:3045–3048
77. (a) Amamoto Y, Aoi Y, Nagashima N, Suto H, Yoshidome D, Arimura Y, Osakabe A, Kato D, Kurumizaka H, Kawashima SA, Yamatsugu K, Kanai M (2017) J Am Chem Soc 139:7568–7576. (b) Ishiguro T, Amamoto Y, Tanabe K, Liu J, Kajino H, Fujimura A, Aoi Y, Osakabe A, Horikoshi N, Kurumizaka H, Yamatsugu K, Kawashima SA, Kanai M (2017) Chem 2:840–859
78. (a) Minoshima M, Bando T, Sasaki S, Fujimoto J, Sugiyama H (2008) Nucleic Acids Res 36:2889–2894. (b) Shinohara K, Yoda N, Takane K, Watanabe T, Fukuyo M, Fujiwara K, Kita K, Nagase H, Nemoto T, Kaneda A (2016) ACS Omega 1:1164–1172
79. (a) Fechter EJ, Dervan PB (2003) J Am Chem Soc 125:8476–8485. (b) Mackay H, Brown T, Sexton JS, Kotecha M, Nguyen B, Wilson WD, Kluza J, Savic B, O'Hare C, Hochhauser D, Lee M, Hartley JA (2008) Bioorg Med Chem 16:2093–2102
80. Trauger JW, Dervan PB (2001) Methods Enzymol 340:450–466
81. (a) Sugiyama H, Hosoda H, Saito I, Asai A, Saito H (1990) Tetrahedron Lett 31:7197–7200. (b) Sugiyama H, Ohmori K, Chan KL, Hosoda M, Asai A, Saito H, Saito I (1993) Tetrahedron Lett 34:2179–2182. (c) Sugiyama H, Fujiwara T, Ura A, Tashiro T, Yamamoto K, Kawanishi S, Saito I (1994) Chem Res Toxicol 7:673–683
82. Pilch DS, Poklar N, Gelfand CA, Law SM, Breslauer KJ, Baird EE, Dervan PB (1996) Proc Natl Acad Sci U S A 93:8306–8311
83. (a) Vashisht Gopal YN, Van Dyke MW (2003) Biochemistry 42:6891–6903. (b) Sunavala-Dossabhoy G, Van Dyke MW (2005) Biochemistry 44:2510–2522
84. (a) Warren CL, Kratochvil NC, Hauschild KE, Foister S, Brezinski ML, Dervan PB, Phillips Jr GN, Ansari AZ (2006) Proc Natl Acad Sci U S A 103:867–872. (b) Puckett JW, Muzikar KA, Tietjen J, Warren CL, Ansari AZ, Dervan PB (2007) J Am Chem Soc 129:12310–12319. (c) Carlson CD, Warren CL, Hauschild KE, Ozers MS, Qadir N, Bhimsaria D, Lee Y, Cerrina F, Ansari AZ (2010) Proc Natl Acad Sci U S A 107:4544–4549
85. (a) Kang JS, Dervan PB (2015) Q Rev Biophys 48:453–464. (b) Anandhakumar C, Li Y, Kizaki S, Pandian GN, Hashiya K, Bando T, Sugiyama H (2014) ChemBioChem 15:2647–2651. (c) Sawatani Y, Kashiwazaki G, Chandran A, Asamitsu S, Guo C, Sato S, Hashiya K, Bando T, Sugiyama H (2016) Bioorg Med Chem 24:3603–3611. (d) Kashiwazaki G, Maeda R, Kawase T, Hashiya K, Bando T, Sugiyama H (2018) Bioorg Med Chem 26:1–7
86. (a) Erwin GS, Bhimsaria D, Eguchi A, Ansari AZ (2014) Angew Chem Int Ed 53:10124–10128. (b) Erwin GS, Grieshop MP, Bhimsaria D, Do TJ, Rodriguez-Martinez JA, Mehta C, Khanna K, Swanson SA, Stewart R, Thomson JA, Ramanathan P, Ansari AZ (2016) Proc Natl Acad Sci U S A 113:7418–7427
87. (a) Chandran A, Syed J, Taylor RD, Kashiwazaki G, Sato S, Hashiya K, Bando T, Sugiyama H (2016) Nucleic Acids Res 44:4014–4024. (b) Chandran A, Syed J, Li Y, Sato S, Bando T, Sugiyama H (2016) ChemBioChem 17:1905–1910

Chapter 2
Development of a New Method for Synthesis of Tandem Hairpin Pyrrole–Imidazole Polyamide Probes Targeting Human Telomeres

Abstract Pyrrole–imidazole (PI) polyamides bind to the minor groove of DNA in a sequence-specific manner without causing denaturation of DNA. To visualize telomeres specifically, tandem hairpin PI polyamides conjugated with a fluorescent dye have been synthesized, but the study of telomeres using these PI polyamides has not been reported because of difficulties synthesizing these tandem hairpin PI polyamides. To synthesize tandem hairpin PI polyamides more easily, we have developed new PI polyamide fragments and have used them as units in Fmoc solid-phase peptide synthesis. Using this new method, we synthesized four fluorescent polyamide probes for the human telomeric repeat TTAGGG, and we examined the binding affinities and specificities of the tandem hairpin PI polyamides, the UV-visible absorption and fluorescence spectra of the fluorescent polyamide probes, and telomere staining in mouse MC12 and human HeLa cells. The polyamides synthesized using the new method successfully targeted to human and mouse telomeres under a mild condition and allow easier labeling of telomeres in the cells while maintaining the telomere structure. Using the fluorescent polyamides, we demonstrated that the telomere length at a single telomere level is related to the abundance of TRF1 protein, a shelterin complex component in the telomere.

Keywords Pyrrole–imidazole polyamides · Tandem hairpin polyamides · Solid-phase peptide synthesis · Telomere · Cellular imaging · TRF1

Y. Kawamoto, *Synthesis and Biological Evaluation of Pyrrole–Imidazole Polyamide Probes for Visualization of Telomeres*, Springer Theses,
https://doi.org/10.1007/978-981-13-6912-4_2

2.1 Introduction

Telomeres locate to the end of chromosomes and are important to the stability and replication of chromosomes [1]. The human telomeric DNA consists of a very long duplex region of the 5′–TTAGGG–3′ sequence with a 3′ overhang and single strands that tend to form G-quadruplexes or t-loops [2]. Vertebrates including human have the repetitive sequence TTAGGG as telomeres [1]. In normal mammalian cells, the number of telomere repeats decreases with cell division, relating to aging process [3, 4]. The telomere length is regulated by a shelterin complex including TRF1, TRF2, Rap1, TIN2, TPP1 and POT1, and is one important biomarker used in various diagnoses [2–5]. Therefore, the design and synthesis of fluorescently labeled polyamides, which target telomeres, have attracted strong interest for visualization and measurement of telomere length.

Dervan and co-workers demonstrated that *N*-methylpyrrole (P)–*N*-methylimidazole (I) (PI) polyamides bind to the minor groove of duplex DNA without causing denaturation of dsDNA and can recognize Watson–Crick base pairs [6]. An antiparallel pairing of I opposite P (I–P) recognizes a G · C base pair, whereas a P–P pair recognizes an A · T or T · A base pair [7]. For efficient binding to long sequences, β-alanine residues are introduced to adjust the pitch of the DNA base pairs and to act as an aliphatic substitution for a P ring [8]. PI polyamides have strong binding affinity and sequence selectivity that can be used for targeting DNA sequences. To date, various structures of PI polyamides have been developed, such as hairpin [9], cyclic [10], and tandem hairpin [11, 12]. These have been used as sequence-specific alkylating agents [13], fluorescent probes [14], and SAHA conjugates [15], and for other uses.

PI polyamides conjugated with fluorescent dyes or alkylating agents have been developed for observing telomeres and studying their influence on the proliferation of cancer cells. Our laboratory has developed fluorescent probes [16] and alkylating agents [17] that target human telomere sequences. Maeshima et al. reported PI polyamide probes that can bind strongly to the human telomeric repetitive sequence TTAGGG ($K_D = 0.51$ nM) in the duplex regions and can be used to stain telomeres in chemically fixed cells [12].

Hairpin PI polyamides with a γ-aminobutyric acid turn (γ-turn) have both sequence specificity and DNA binding affinity [9], and they can be synthesized easily by machine-assisted Fmoc solid-phase peptide synthesis (SPPS) [18]. Tandem hairpin PI polyamides comprise two hairpin PI polyamides linked with a flexible linker and can recognize ≥10 base pairs [11, 12]. However, the yield of tandem hairpin PI polyamides was low because multiple steps were needed to remove Boc groups or Fmoc groups in Fmoc SPPS or Boc SPPS, respectively. Therefore, the interruption of automatic coupling reactions for the deprotection of amino groups was needed [11a, b, d]. The detailed synthetic protocols for some polyamides have not been described [12].

To overcome the difficulties associated with the synthesis of tandem hairpin PI polyamides, we herein demonstrate a new synthetic route for creating tandem hairpin PI polyamides in which the tetraamide unit, **9**, synthesized in solution phase, is introduced into Fmoc SPPS. Using this method, we synthesized four polyamide–fluorescent probes conjugated with Texas Red (TR) or Cyanine 3 (Cy3) to visualize mammalian telomeres. This new synthetic route is more efficient and applicable to a wider variety of tandem hairpin PI polyamides compared with the methods reported previously because machine-assisted SPPS without any interruptions can provide tandem hairpin PI polyamides. Thermal denaturation analysis ensured their binding to and specificities for human telomeric repetitive sequence. The synthesized fluorescent polyamide probe conjugates were evaluated by measuring the UV-visible absorption and fluorescence intensity and by telomere staining of the polyamides in mouse and human cells. This new method simplifies the synthesis of tandem hairpin PI polyamides and also the observation of human telomeres using these PI polyamide–fluorescent probes.

2.2 Results and Discussion

2.2.1 Synthesis of Conjugates 1–4

To synthesize the conjugates **1–4** (Fig. 2.1), we synthesized the new tetraamide unit **9** (Scheme 2.1). Reduction of the compound **5**, synthesized using the previous method [13a], was performed under high pressure (0.23 MPa) of hydrogen gas with 10% Pd/C to afford amine crude powder, which was coupled with 1-methyl-1*H*-imidazole-2-carboxylic acid using benzotriazol-1-yl-oxytripyrrolidinophosphonium hexafluorophosphate (PyBOP) and *N,N*-diisopropylethylamine (DIEA) in *N,N*-dimethylformamide (DMF) to produce the compound **6**. The crude compound **6** was hydrolyzed to obtain the first key compound **7**. Next, the crude compound **7** was activated with PyBOP to afford the compound **8**, followed by coupling with the deblocked Fmoc-D-Dab(Boc)-OH to produce the second key compound **9**. The crude compound **9** was washed with diethyl ether and then triturated with water. Using the compounds **7** and **9** as units without further purification, machine-assisted solid-phase peptide synthesis shown in Scheme 2.2 was performed to afford the tandem hairpin PI polyamides **10** (named **TH59** [12]) and **11** (named **Deamino TH59**). The compound **11**, which does not have an amino group at the γ-turn, was designed for the ease of synthesis of tandem hairpin PI polyamides. The new units **7** and **9** allowed us to omit the steps to remove Boc groups on the resin and to synthesize tandem hairpin PI polyamides automatically and easily. The crude tandem hairpin PI polyamides **10** and **11** were cleaved from the resin using 3, 3′-diamino-*N*-methyldipropylamine and then purified by reversed-phase flash chromatography, followed by coupling with TR succinimidyl ester or Cy3-NHS ester in DMF and DIEA to afford conjugates **1–4** (**TH59TR**, **Deamino TH59TR**, **TH59Cy3**, **Deamino TH59Cy3**, respectively) after purification by reversed-phase HPLC (Scheme 2.2). ESI-TOF MS spectra of **1–4**, **10**, and **11** are shown in Figs. S2.1 and S2.2.

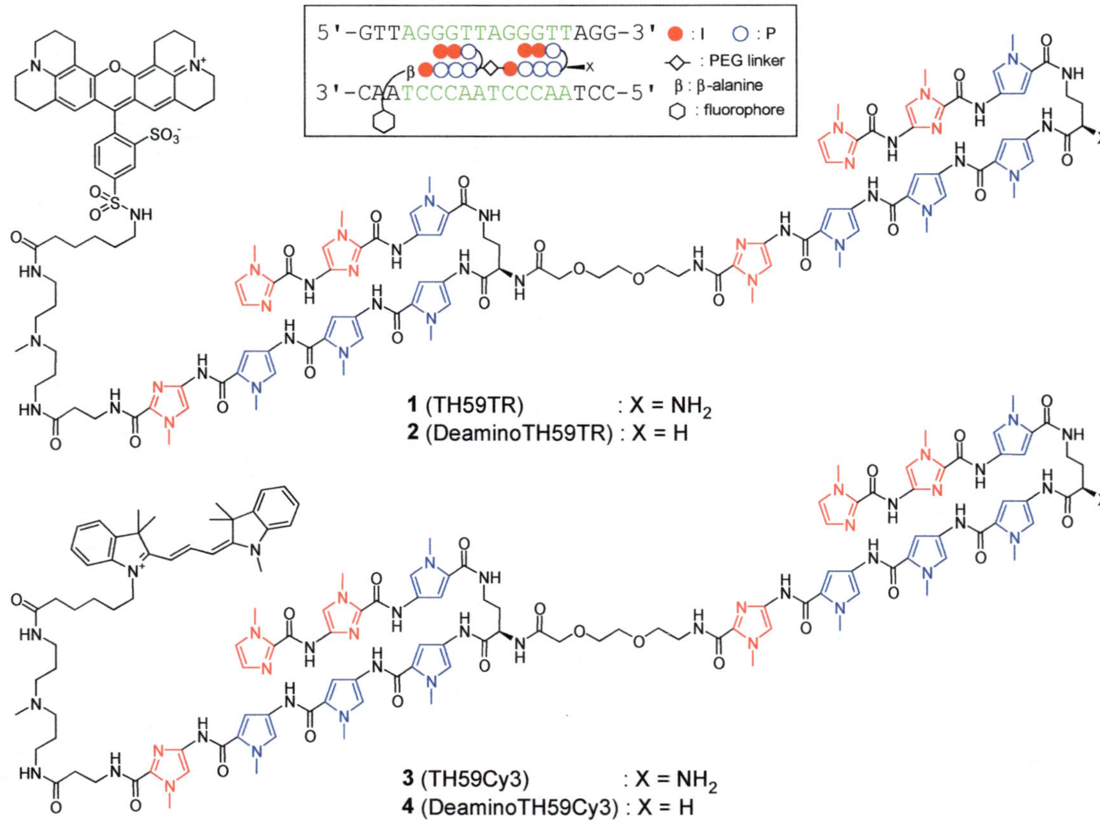

1 (TH59TR) : X = NH_2
2 (DeaminoTH59TR) : X = H

3 (TH59Cy3) : X = NH_2
4 (DeaminoTH59Cy3) : X = H

Fig. 2.1 Chemical structure of tandem hairpin fluorescent PI polyamides **1–4** targeting the human telomere sequence and their ball-and-stick representation.

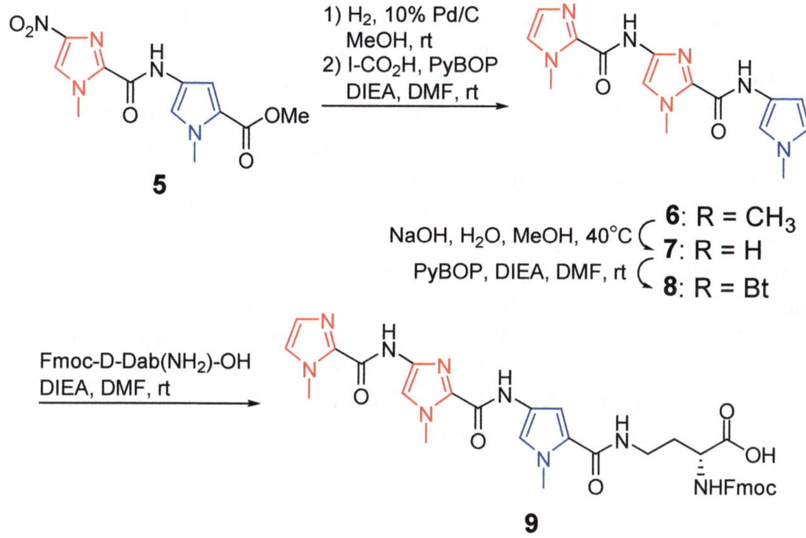

Scheme 2.1 Synthesis of PI polyamide Units **7** and **9** for solid-phase synthesis. Copyright 2013 American Chemical Society

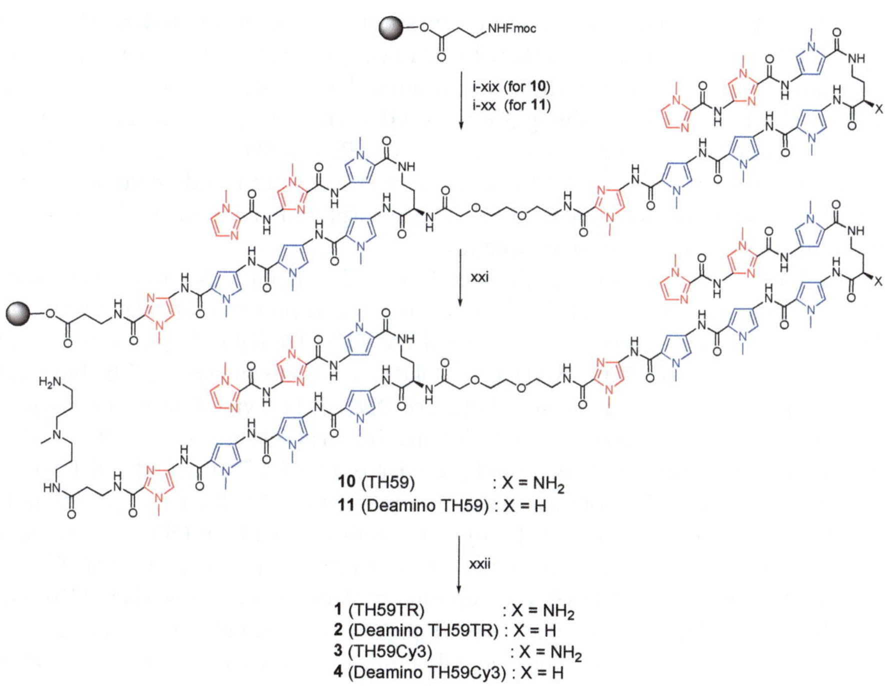

Scheme 2.2 Solid-phase Synthesis of Tandem Hairpin Fluorescent PI Polyamides **1–4**

All HCTU-mediated coupling conditions were performed under similar conditions (see ref. [10c]). Reagents and conditions: (i) 20% piperidine, DMF; (ii) Fmoc-PI-CO_2H, HCTU, DIEA, DMF; (iii) 20% piperidine, DMF; (iv) Fmoc-P-CO_2H, HCTU, DIEA, DMF; (v) 20% piperidine, DMF; (vi) Fmoc-P-CO_2H, HCTU, DIEA, DMF; (vii) 20% piperidine, DMF; (viii) **9**, HCTU, DIEA, DMF; (ix) 20% piperidine, DMF; (x) Fmoc-mini-PEG-OH, HCTU, DIEA, DMF; (xi) 20% piperidine, DMF; (xii) Fmoc-PI-CO_2H, HCTU, DIEA, DMF; (xiii) 20% piperidine, DMF; (xiv) Fmoc-P-CO_2H, HCTU, DIEA, DMF; (xv) 20% piperidine, DMF; (xvi) Fmoc-P-CO_2H, HCTU, DIEA, DMF; (xvii) 20% piperidine, DMF; (xviii) **9**, HCTU, DIEA, DMF (for **10**) or Fmoc-γ-Abu-OH, HCTU, DIEA, DMF (for **11**); (xix) 20% piperidine, DMF; (xx) **7**, PyBOP, DIEA, DMF (for **11**); (xxi) 3,3′-diamino-*N*-methyldipropylamine, 45 °C; (xxii) TR succinimidyl ester, DIEA, DMF (for **1** and **2** from **10** and **11**, respectively) or Cy3-NHS ester, DIEA, DMF (for **3** and **4** from **10** and **11**, respectively). Copyright 2013 American Chemical Society.

2.2.2 UV-Visible Absorption Spectra and Fluorescence Spectra

The UV-visible absorption and fluorescence spectra of conjugates **1–4** are shown in Fig. 2.2. The absorption spectra (black lines) of conjugates **1–4** were measured at the same concentration (6 μM) in 5 mM Na phosphate buffer (pH 7.0; **1** and **3** containing 10% v/v DMF, **2** and **4** containing 3% v/v DMF). The absorption maximum due to the dye of **1** and **2**, which were conjugated with TR, was 596 nm, and that of **3** and **4**, conjugated with Cy3, was 549 nm. These maxima were mostly consistent with each dye's excitation wavelength [14a]. The absorption maximum at ~310 nm was due to the pyrrole and imidazole moieties.

The fluorescence spectra of conjugates **1–4** (3 μM) in 5 mM Na phosphate buffer (pH 7.0 containing 5% v/v DMF) in the absence and presence of oligonucleotide (ODN)-1/2, whose sequences were 5′-GGTTAGGGTTAGGGTTAGG-3′ (ODN-1) and 3′-CCAATCCCAATCCCAATCC-5′ (ODN-2), are shown in Fig. 2.2 by blue and red lines, respectively. In each case, addition of ODN-1/2 provided stronger emission, but compounds **1** and **2** having an emission maximum at 609 nm ($\lambda_{ex} = 596$ nm) differed slightly between the absence and presence of ODN-1/2 (1.3-fold and 1.1-fold greater, respectively). By contrast, the emission maxima at 560 nm ($\lambda_{ex} = 549$ nm) of **3** and **4** were 3.0-fold and 3.9-fold greater with than without ODN-1/2, respectively. Similar enhancement of the fluorescence of tetramethylrhodamine (TMR) and perylene by addition of dsDNA was reported by Dervan and co-workers [19] and our laboratory [16], respectively. It was suggested that hydrophobic intramolecular interactions between the TMR and the polyamide formed in the absence of dsDNA would induce electron transfer from the polyamide to the singlet excited state of the TMR and the fluorescence would be quenched, and this interaction would be disrupted in the presence of matched dsDNA, resulting in an increase in fluorescence [19b]. The changes in fluorescence of the polyamide–Cy3 conjugates may occur

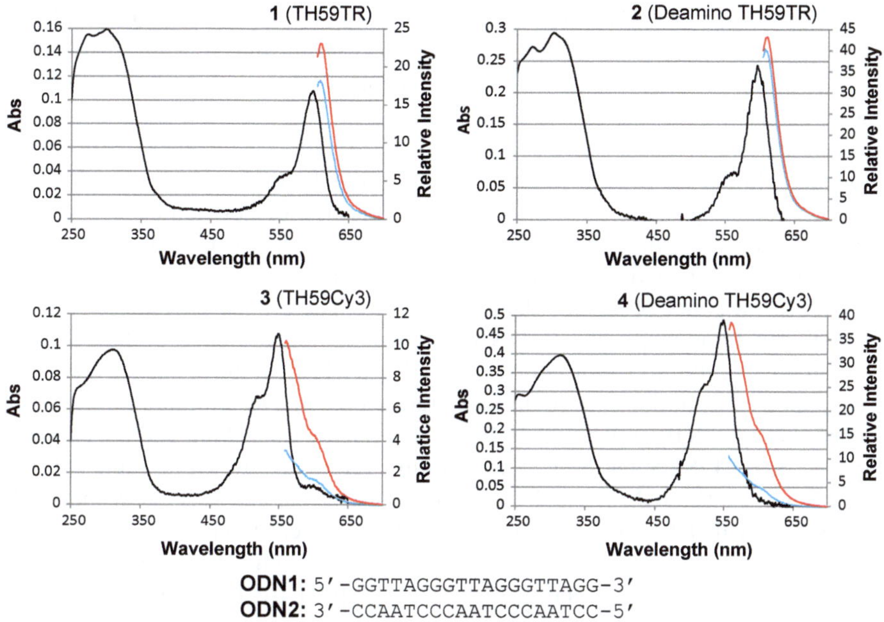

Fig. 2.2 UV-visible absorption spectra (black) of conjugates **1–4** ([**1**] = [**2**] = [**3**] = [**4**] = 6 μM) in a 5 mM Na phosphate buffer (pH 7.0, **1** and **3** containing 10% v/v DMF, **2** and **4** containing 3% v/v DMF) and fluorescence emission spectra of conjugates **1–4** (3 μM) in the absence of ODN-1/2 (blue) or presence of 6 μM ODN-1/2 (red) in 5 mM Na phosphate buffer (pH 7.0 containing 5% v/v DMF). Copyright 2013 American Chemical Society

presumably because of this quenching, and therefore this system may not be applied to the polyamide–TR conjugates.

2.2.3 Binding Affinities and Specificities of Tandem Hairpin PI Polyamides

Thermal stabilization of polyamide-DNA complex can be analyzed by thermal melting temperature (T_m) analysis and this method has been used for the measurement of the relative binding affinity and the ability to discriminate mismatch sequences [20]. In order to confirm whether tandem hairpin PI polyamides **10** (**TH59**) and **11** (**Deamino TH59**) bind to telomeric repeats strongly and specifically, thermal denaturation analysis of DNA and of polyamides-DNA complex was performed. The sequences of dsDNA were 5′-GGTTAGGGTTAGGGTTAGG-3′ (ODN-1) and 3′-CCAATCCCAA TCCCAATCC-5′ (ODN-2), 5′-GGTTAGAGTTAGGGTTAGG-3′ (ODN-3) and 3′-CCAATCTCAATCCCAATCC-5′ (ODN-4), 5′-GGTT AGGGTTAGAGTTAGG-3′ (ODN-5) and 3′-CCAATCCCAATC TCAATCC-

5′ (ODN-6), and 5′-GGTTAGAGTTAG AGTTAGG-3′ (ODN-7) and 3′-CCAA TCTCAATCTCAATCC-5′ (ODN-8), which had no mismatch, 1 bp mismatch at the side of the polyamides' C-termini, 1 bp mismatch at the side of their N-termini, and 2 bp mismatch, respectively. The underlined bases are the binding sites of polyamides and the bold bases show mismatched parts. The analysis buffer is the aqueous solution of 10 mM sodium chloride and 10 mM sodium cacodylate at pH 7.0 containing 2.5% v/v DMF. The concentration of polyamides and dsDNA was 12.5 and 2.5 μM, respectively (5:1 stoichiometry). When analyzing in other stoichiometry (1:1, and 2:1), reasonable melting transition at a stoichiometry of 1:1 and 2:1 was not observed (data not shown). There is an example of the UV thermal denaturation profile of DNA duplexes in the presence of DNA binder targeting 10–12 bp at 5 mol equivalent or higher [21]. Before analyses, samples were annealed from 80 °C to room temperature for 1 h. Denaturation profiles were recorded at λ = 260 nm from 25 to 90 °C at a rate of 1.0 °C/min and are shown in Figs. S2.3, S2.4, S2.5 and S2.6. The resulting values are shown in Table 2.1. The ΔT_m values of **10**-ODN-1/2 and **11**-ODN-1/2 were 26.0 and 19.1 °C, respectively, demonstrating that these tandem hairpin PI polyamides can bind to TTAGGG repeats and, especially, the affinity of **10** was equivalent to cyclic PI polyamides, which have high binding affinities (i.e. $\Delta T_m > 20.0$ °C) [10a, d]. The compound **10** have the higher affinity than **11** presumably because of the amino group at the γ-turn [22]. ΔT_m(match)-ΔT_m(mismatch) ($\Delta\Delta T_m$) values of **10**-ODN-3/4 and **11**-ODN-3/4 were 7.6 and 6.9 °C, respectively. In contrast, $\Delta\Delta T_m$ values of **10**-ODN-5/6 and **11**-ODN-5/6 were 7.2 and 6.1 °C, respectively. Moreover the complex of **10** and **11** and ODN-7/8, which has 2 bp mismatch, show higher $\Delta\Delta T_m$ values, 14.3 and 13.3 °C, respectively. These results suggest tandem hairpin PI polyamides can discriminate 1 and 2 bp mismatch and the hairpin PI polyamides at the side of C-termini have slightly better abilities of the discrimination than those at the side of N-termini.

Table 2.1 T_m values for tandem hairpin PI polyamides **10** and **11**

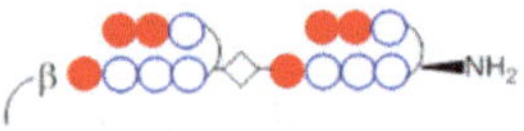

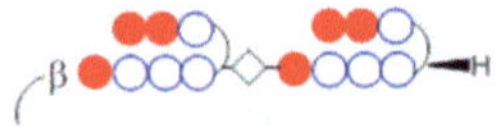

	10 (TH59)			**11** (Deamino TH59)		
dsDNA sequences targeted by polyamides	T_m	ΔT_m	ΔT_m(match)-ΔT_m(mismatch)	T_m	ΔT_m	ΔT_m(match)-ΔT_m(mismatch)
Match DNA (T_m : 43.7°C (±0.4°C)) 5′- GGTT**AGGGTTAGGGTT**AGG -3′ 3′- CCAA**TCCCAATCCCAA**TCC -5′	69.7°C (±0.5°C)	26.0°C	—	62.8°C (±0.3°C)	19.1°C	—
1 bp Mismatch DNA (T_m : 44.5°C (±0.5°C)) 5′- GGTT**AGAGTTAGGGTT**AGG -3′ 3′- CCAA**TCTCAATCCCAA**TCC -5′	62.8°C (±0.2°C)	18.3°C	7.6°C	56.7°C (±0.7°C)	12.2°C	6.9°C
1 bp Mismatch DNA (T_m : 44.6°C (±0.5°C)) 5′- GGTT**AGGGTTAGAGTT**AGG -3′ 3′- CCAA**TCCCAATCTCAA**TCC -5′	63.3°C (±0.2°C)	18.8°C	7.2°C	57.6°C (±0.2°C)	13.0°C	6.1°C
2 bp Mismatch DNA (T_m : 43.4°C (±0.3°C)) 5′- GGTT**AGAGTTAGAGTT**AGG -3′ 3′- CCAA**TCTCAATCTCAA**TCC -5′	55.1°C (±0.3°C)	11.7°C	14.3°C	49.3°C (±0.3°C)	5.9°C	13.3°C

All T_m values are the average calculated from at least 4 melting temperature analyses with standard deviations indicated in parentheses. ΔT_m values are given as T_m(polyamide-DNA complex)—T_m(DNA). Copyright 2013 American Chemical Society

2.2.3.1 Telomere Targeting of the Synthesized Fluorescent Polyamides in Mouse and Human Cells

We doubly stained mouse MC12 cell spreads with the fluorescent polyamides and 4′, 6-diamidino-2-phenylindole (DAPI) to compare the ability of conjugates **1–4** (**TH59TR**, **Deamino TH59TR**, **TH59Cy3**, **Deamino TH59Cy3**) to stain telomeres specifically. We examined various staining conditions and found that the optimal result was obtained using PI polyamide binding with low-salt washing, followed by higher-salt washing, as described in the Materials and Methods. The images are shown in Fig. 2.3. The DAPI signal shows chromosomal regions. The PI polyamide signals show intense foci, which represent the subchromosomal binding with the polyamides. In each case, two foci are observed at every chromosomal end, suggesting that all **TH59TR**, **Deamino TH59TR**, **TH59Cy3**, and **Deamino TH59Cy3** can bind to the mouse telomeric repeat TTAGGG. The background signals increased when stained with **TH59Cy3** and **Deamino TH59Cy3**, whereas little background was detected along chromosomes treated with the **TH59TR** and **Deamino TH59TR**. This result is presumably because Cy3 induced nonspecific binding of polyamides, but TR did not induce such binding. Although in vitro analysis showed that **TH59Cy3** and **Deamino TH59Cy3** fluoresced more strongly when bound to the targeted DNA (shown in Fig. 2.2) and should have distinguished telomeric repeats, the cell staining image suggested that TR conjugates are superior to Cy3 conjugates for observing telomeres in the cells. Furthermore, PI polyamide signals in **TH59TR** images were slightly stronger than those in **Deamino TH59TR** images, presumably because, shown in Table 2.1 as T_m values of not-conjugated polyamides, the former has higher affinities to human telomeric repeats than the latter.

Next, following the previous report [12], we used immunofluorescence to prove that the **TH59TR** and **Deamino TH59TR** can localize to telomeres. We reasoned that a shelterin complex component in telomeres, TRF1 protein, which binds to TTAGGG repeats and controls telomere length [2–5], should colocalize with the **TH59TR** and **Deamino TH59TR**. Figure 2.4 shows human HeLa cells treated with the **TH59TR** and **Deamino TH59TR** (red) together with the immunofluorescence signal of TRF1 antibodies (green). This examination suggested that the signals of **TH59TR** and **Deamino TH59TR** were colocalized with that of TRF1 because the resulting overlapping yellow signals were observed.

Furthermore, **TH59TR** and TRF1 signals were extracted from the Fig. 2.4a image and their intensities were compared at a single telomere level. As shown in Fig. 2.5, we found a close correlation at a single telomere level between signals of **TH59TR** and TRF1-antibody (correlation coefficient: 0.80), suggesting telomere length is highly related to abundance of shelterin complex in telomeres. This finding is a good agreement with the "protein-counting" model of telomere length regulation [2–5]. Various methods of telomere length analysis have been developed, but they require denaturation or electrophoresis steps [23]. These observations demonstrate that **TH59TR** and **Deamino TH59TR** are telomere-specific polyamide probes that allows us to observe human telomeres without breaking the telomere structures and to estimate telomere length easily with a simple staining protocol.

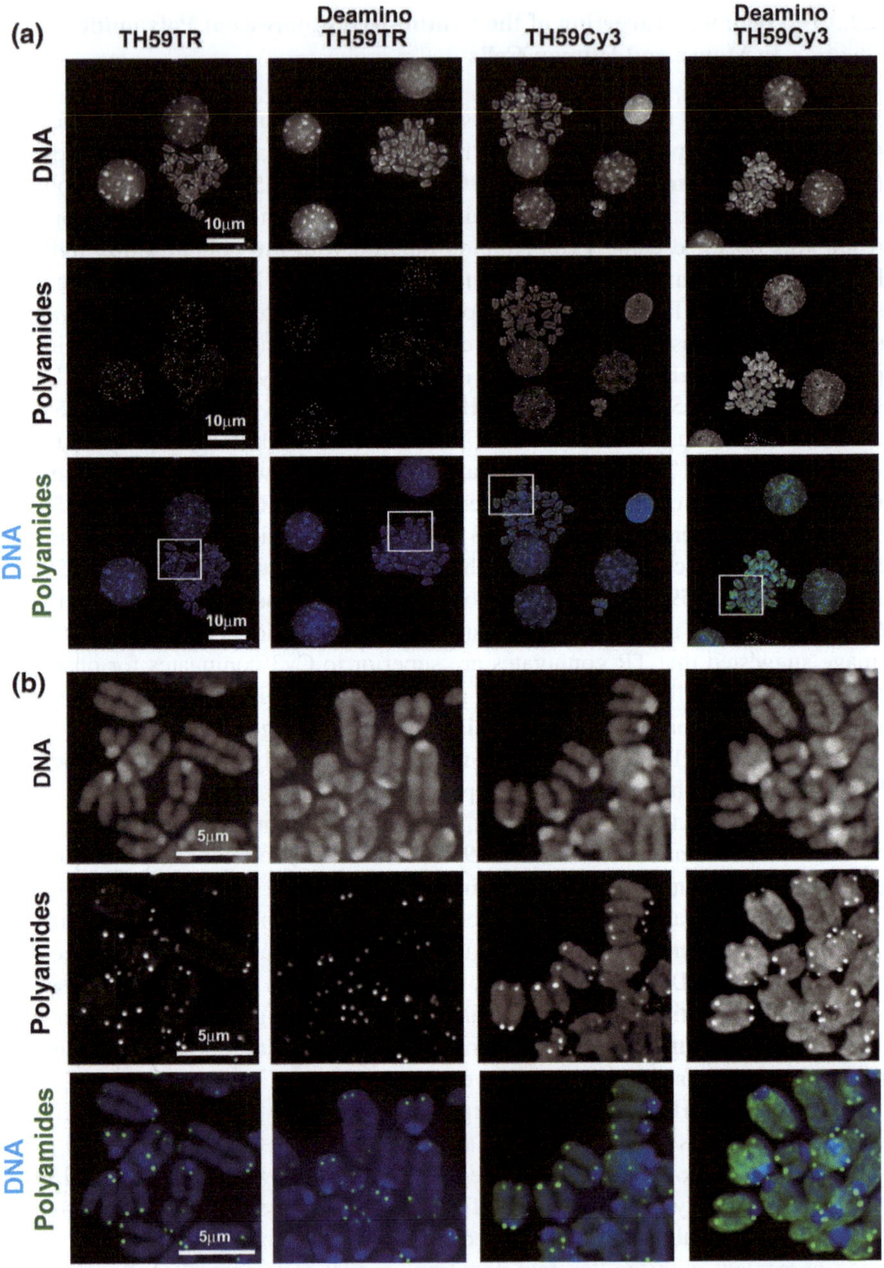

Fig. 2.3 Telomere Staining of Mouse MC12 Cell Spreads with Fluorescent Polyamides. **a** The MC12 spreads were stained with DAPI (1st row) and the fluorescent polyamides (2nd row): 1st column, **TH59TR**; 2nd column, **Deamino TH59TR**; 3rd column, **TH59Cy3**; 4th column, **Deamino TH59Cy3**. The merged images of DAPI and polyamides staining are shown on 3rd row. Enlarged images of the boxed region in panel (**a**) are demonstrated in panel (**b**). Copyright 2013 American Chemical Society

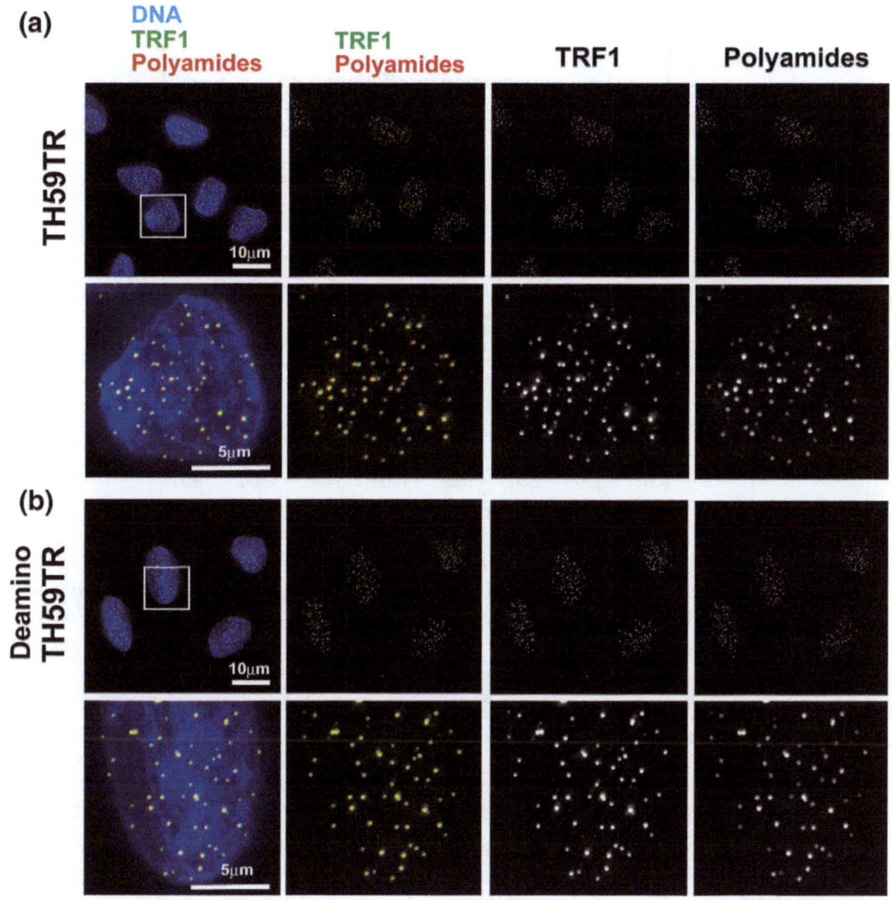

Fig. 2.4 Staining of telomeric repeats with fluorescent polyamides and co-localization with telomere binding protein TRF1. HeLa1.3 cells were stained with DAPI, TRF1 antibody and **TH59TR** (**a**) or **Deamino TH59TR** (**b**). Enlarged images of the boxed regions in 1st row images are shown on 2nd row. Note that **TH59TR** or **Deamino TH59TR** foci co-localize with immunofluorescence signals with TRF1 antibodies. Copyright 2013 American Chemical Society

Moreover, we examined the effect of the washing process on telomere staining. We detected the telomeric signals without a washing process; however, this process helped reduce the background or nonspecific PI polyamide binding (shown in Fig. S2.7).

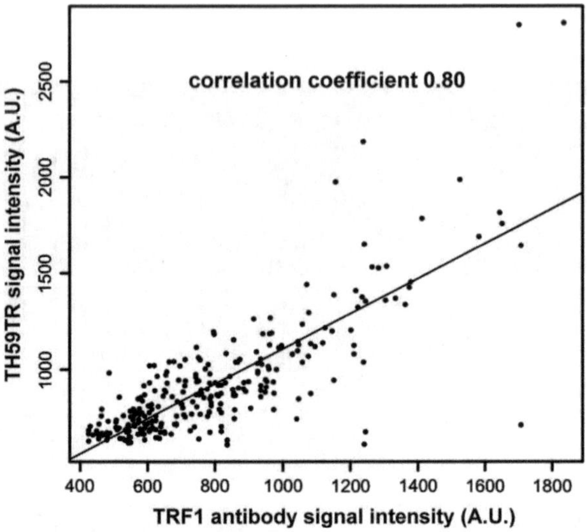

Fig. 2.5 Correlation between Signal Intensities of THd59TR and Immunofluorescence for TRF1. **TH59TR** and TRF1 signals were extracted from the Fig. 2.4a image and their intensities were compared at single telomere level and shown as a scattered plot. A.U., arbitrary unit. Copyright 2013 American Chemical Society

2.3 Conclusion

In this study, we developed a synthetic scheme for tandem hairpin PI polyamides with two units of Fmoc SPPS, which we used to synthesize telomere-targeting polyamides conjugated with TR or Cy3. In mouse and human cells, TR-conjugated polyamides **1** and **2 (TH59TR** and **Deamino TH59TR**, respectively) highlighted the telomere foci clearly. This novel method simplifies the synthesis of PI polyamide probes targeting the human telomeric repeat TTAGGG and is useful for methods to label and analyze telomeres in the cells without breaking the telomere structures.

2.4 Supporting Information

2.4.1 Materials for the Synthesis

^{1}H NMR spectra were recorded on a JEOL JNM ECA-600 spectrometer (600 MHz for ^{1}H), with chemical shifts reported in parts per million relative to residual solvent and coupling constants in hertz. The following abbreviations were applied to spin multiplicity: s (singlet), d (doublet), t (triplet), q (quartet), and m (multiplet). HPLC analysis was performed on a Jasco Engineering PU-2089 plus series system using

4.6 × 150 mm X-Terra MS C_{18} reversed-phase column in 0.1% TFA in water with acetonitrile as the eluent at a flow rate of 1.0 mL/min and a linear gradient elution of 0–100% acetonitrile in 40 min with detection at 254 nm. HPLC purification was performed with a Jasco Engineering UV2075 HPLC UV/VIS detector and a PU-2080 plus series system using a 4.6 × 150 mm Chemcobond 5-ODS-H reversed-phase column in 0.1% TFA in water with acetonitrile as the eluent at a flow rate of 1.0 mL/min and a linear gradient elution of 30–75% acetonitrile in 20 min with detection at 254 nm. Collected fractions were analyzed by ESI-TOF-MS (Bruker). For the collect mass measurement, external mass calibration (m/z: 622.0290 and 922.0098), purchased from Agilent Technologies, was used. Reversed-phase flash chromatography was performed on CombiFlash Rf (Teledyne Isco, Inc.) using a 4.3 g reversed-phase flash column (C_{18} RediSep Rf) in 0.1% TFA in water with acetonitrile as the eluent at a flow rate of 18.0 mL/min and a linear gradient elution of 0–35% acetonitrile in 5–30 min with detection at 254 nm. DNA oligonucleotides were purchased from Sigma. Fmoc-β-Ala-Wang resin (0.55 mmol/g), Fmoc-mini-PEG-OH, and Fmoc-D-Dab(Boc)-OH were purchased from Peptide International. Fmoc-P-CO_2H, Fmoc-PI-CO_2H, DMF used for solid-phase synthesis, 1-methyl-2-pyrrolidone (NMP), trifluoroacetic acid (TFA), and piperidine were from Wako. O-(1H-6-chlorobenzotriazol-1-yl)-1, 1, 3, 3-tetramethyluronium hexafluorophosphate (HCTU) was from Peptide Institute, Inc. Fmoc-β-Ala-Wang resin, Fmoc-γ-Abu-OH, and PyBOP were from Novabiochem. DIEA was from Nacalai Tesque, Inc. DMF used for solution-phase synthesis was from Kanto Chemical Co., Inc. 3,3′-diamino-*N*-methyldipropylamine, 10% Pd/C, and 1-methyl-1*H*-imidazole-2-carboxylic acid were from Aldrich. TR succinimidyl ester and Cy3-NHS ester were from Molecular Probes. All other solvents and materials were from standard suppliers (highest quality available).

2.4.2 Materials for the Analysis of Spectra and Thermal Denaturation Analysis

Polyamide concentrations were calculated with a Nanodrop ND-1000 spectrophotometer (Thermo Fisher Scientific Inc.) using an extinction coefficient of 9900 $M^{-1}cm^{-1}$ per one pyrrole or imidazole moiety at λ_{max} near 310 nm. UV-Vis spectra were measured on a Spectrophotometer V-650 (JASCO). Steady-state fluorescence spectra were obtained on a Spectrofluorometer FP-6300 (JASCO). These two kinds of spectra were obtained at room temperature. Melting temperature analyses were performed on a Spectrophotometer V-650 (JASCO) equipped with a thermo-controlled PAC-743R cell changer (JASCO) and a refrigerated and heating circulator F25-ED (Julabo).

2.4.3 Synthetic Details of PI Polyamides

Synthesis of IIP-CO_2Me (6). The compound **6** was synthesized from O_2N-IP-CO_2Me (**5**). To a solution of **5** (2001 mg, 6.51 mmol) in MeOH (50 mL) was added 10% Pd/C (100 mg), and the reaction mixture was stirred for 6 h at room temperature under high pressure (0.23 MPa) of hydrogen gas. The reaction mixture was filtered through the filter paper with dichloromethane and concentrated in vacuo to afford 1478 mg of yellow amine crude, which was used in the next coupling without further purification. To a solution of the amine in DMF 25 mL was added a solution of 1-methyl-1*H*-imidazole-2-carboxylic acid (651 mg, 5.16 mmol) and PyBOP (2685 mg, 5.16 mmol) in DIEA (1.89 mL, 10.9 mmol) and DMF 30 mL and the reaction mixture was stirred over weekend. After concentration, the residue was triturated with water as a brown solid. The residue was dissolved in dichloromethane and concentrated in vacuo to afford **6** (1931 mg, 77% yield for 2 steps). The compound was used in the next reaction without further purification. Analytical HPLC: t_R = 16.0 min. ESI-TOF-MS *m/z* calcd for $C_{17}H_{20}N_7O_4^+$ $[M + H]^+$ 386.1571 found 386.1422. ^{1}H NMR (600 MHz, DMSO-d_6) δ 10.49 (s, 1H; NH), 9.78 (s, 1H; NH), 7.57 (s, 1H; CH), 7.53 (d, 1H, J = 1.3 Hz; CH), 7.46 (s, 1H; CH), 7.08 (s, 1H; CH), 7.01 (d, 1H, J = 1.3 Hz; CH), 4.01 (s, 3H; NMe), 4.00 (s, 3H; NMe), 3.83 (s, 3H; NMe), 3.74 (s, 3H; OMe).

Synthesis of IIP-CO_2H (7). The compound **7** was synthesized from IIP-CO_2Me (**6**) by hydrolysis. To a solution of **6** (1812 mg, 4.73 mmol) in methanol/water (1:1, total 40 mL) was added sodium hydroxide (2.5 g) and the reaction mixture was stirred for 6 h at 40 °C, as followed by the evaporation of methanol. The base was neutralized by the addition of hydrochloric acid at 0 °C to afford precipitations. The precipitations were dried in vacuo to afford the compound **7** (1102 mg, 60% yield) as a brown powder. The compound was used in the next coupling without further purification. Analytical HPLC: t_R = 13.5 min. ESI-TOF-MS *m/z* calcd for $C_{16}H_{18}N_7O_4^+$ $[M + H]^+$ 372.1415 found 372.1440. ^{1}H NMR (600 MHz, DMSO-d_6) δ 10.41 (s, 1H; NH), 9.86 (s, 1H; NH), 7.57 (s, 1H; CH), 7.48 (s, 1H; CH), 7.47 (d, 1H, J = 2.1 Hz; CH), 7.12 (s, 1H; CH), 6.96 (d, 1H, J = 2.0 Hz; CH), 4.01 (s, 3H; NMe), 4.00 (s, 3H; NMe), 3.83 (s, 3H; NMe).

Synthesis of Fmoc-D-Dab(IIP)-OH (9). Compound **9** was synthesized from **7** and Fmoc-D-Dab(Boc)-OH. The procedure for the synthesis is as follows. To remove the Boc group, TFA (487.5 μL) and water (12.5 μL) was added to Dab (100 mg, 0.23 mmol) and mixed until Dab dissolved. After concentration of the mixture, the residue was washed with diethyl ether and triturated with dichloromethane-diethyl ether, then dried in vacuo. DMF (1 mL) and DIEA (200 μL) was added to the resulting Fmoc-D-Dab(NH_2)-OH as a white powder. On the other hand, crude **7** (73 mg, 0.20 mmol) and PyBOP (103 mg, 0.20 mmol) were dissolved in DMF (500 μL) and DIEA (160 μL) and mixed to afford the activated compound **8**. At room temperature, the solution of the compound **8** was added slowly to Fmoc-D-Dab(NH_2)-OH solution and the reaction solution was mixed for 10 min and then settled for 1 h at room temperature. After concentration in vacuo, the residue was washed with diethyl ether and then triturated with water. The resulting residue was dried to allow the compound **9** (134 mg, 98% yield) as a yellow powder. The compound was used in the next solid-phase synthesis without purification. Analytical HPLC: $t_R = 21.3$ min. ESI-TOF-MS *m/z* calcd for $C_{35}H_{36}N_9O_7^+$ $[M + H]^+$ 694.2732 found 694.2723. ^1H NMR (600 MHz, DMSO-d_6), δ 10.28 (s, 1H; NH), 9.74 (s, 1H; NH), 8.06 (t, 1H, J = 5.5 Hz; NH), 7.90 (d, 2H, J = 8.3 Hz; CH × 2), 7.74 (t, 2H, J = 6.5 Hz; CH × 2), 7.72 (d, 1H, J = 8.2 Hz; NH), 7.56 (s, 1H; CH), 7.46 (s, 1H; CH), 7.42 (t, 2H, J = 7.6 Hz; CH × 2), 7.33 (t, 2H, J = 7.6 Hz; CH × 2), 7.22 (d, 1H, J = 1.4 Hz; CH), 7.08 (s, 1H, CH), 6.98 (d, 1H, J = 1.4 Hz; CH), 4.28–4.29 (m, 2H; CH_2), 4.23–4.26 (m, 1H; CH), 4.02–4.05 (m, 1H; CH), 4.01 (s, 3H; NMe), 4.00 (s, 3H; NMe), 3.80 (s, 3H; NMe), 2.89–3.08 (m, 2H; CH_2), 1.99–2.04 (m, 1H; CH), 1.79–1.85 (m, 1H; CH).

General Procedures of Fmoc Solid-phase Synthesis. Each synthesis of tandem hairpin PI polyamides (**10, 11**) was performed on a PSSM-8 (Shimadzu) with a computer-assisted operation system on a 0.02 mmol scale (about 40 mg of Fmoc-β-Wang resin, 0.55 mmol/g) by using Fmoc chemistry. An Fmoc unit (0.20 mmol) in each step was set up to solve by NMP (1 mL) on the synthetic line. Fmoc units in each step were as follows: Fmoc-P-CO_2H (77 mg), Fmoc-PI-CO_2H (70 mg), IIP-CO_2H (70 mg) (**7**), Fmoc-D-Dab(IIP)-OH (70 mg) (**9**), Fmoc-mini-PEG-OH (81 mg), and Fmoc-γ-Abu-OH (69 mg). Procedures were as follows: twice deblocking for 4 min with 20% piperidine/DMF (0.5 mL), activating for 2 min with HCTU (88 mg, 0.21 mmol) in DMF (1 mL) and 10% DIEA/DMF (0.4 mL), coupling for 60 min, and washing with DMF. At the last coupling of **11**, a solution of the activated compound **7** (70 mg, 0.19 mmol) with PyBOP (103 mg, 0.20 mmol) in 18% DIEA/NMP 0.91 mL was used. After the last coupling of **10**, the deblocking of amino group at γ-turn was performed. All coupling were carried out with a single-coupling cycle. All lines were purged with solution transfers and bubbled by nitrogen gas for stirring the resin.

Synthesis of IIP-D-Dab-PPPI-PEG-D-Dab(IIP)-PPPI-β-amine (10, TH59). Solid-phase synthesis was performed with 44 mg resin. After that, the compound supported by Wang resin was cleaved using 3,3′-diamino-*N*-methyldipropylamine (96% assay). 0.4 mL of that was added, followed by stirring for 3 h at 45 °C. The resin was removed by filtration and washed thoroughly with dichloromethane, and the filtrate was concentrated in vacuo. The residue was washed with diethyl ether and triturated with dichloromethane-diethyl ether, then dried in vacuo. The resulting crude polyamide (32.0 mg) was purified by reversed-phase flash chromatography and appropriate fractions were collected under freeze-dry condition to give the compound **10** (3.7 mg, 1.6×10^{-3} mmol, 6.9% yield in 10 steps). Analytical HPLC: $t_R = 14.9$ min. ESI-TOF-MS *m/z* calcd for $C_{102}H_{130}N_{41}O_{20}^{3+}$ $[M + 3H]^{3+}$ 749.6800 found 749.6642.

Synthesis of IIP-γ-PPPI-PEG-D-Dab(IIP)-PPPI-β-amine (11, Deamino TH59). Solid-phase synthesis was performed with 44 mg resin. After solid-phase synthesis and similar cleavage, the crude compound **11** was gained according to the above procedure. The resulting crude polyamide (22.0 mg) was purified similarly to give the compound **11** (5.5 mg, 2.5×10^{-3} mmol, 10% yield in 11 steps). Analytical

HPLC: $t_R = 16.0$ min. ESI-TOF-MS *m/z* calcd for $C_{102}H_{129}N_{40}O_{20}{}^{3+}$ $[M + 3H]^{3+}$ 744.6763 found 744.6678.

Synthesis of the Conjugate 1 (TH59TR). First TR succinimidyl ester (1.0 mg) in DMF (80 μL) was prepared. The compound **10** (0.8 mg, 3.6×10^{-4} mmol) was dissolved in DMF (199.8 μL) and DIEA (0.2 μL), and the solution of TR succinimidyl ester (48 μL, 7.4×10^{-4} mmol) was added to that of **10**, followed by mixing for 4 h at room temperature with shielding the light. The reaction mixture was purified by reversed-phase HPLC and appropriate fractions were collected under freeze-dry conditions to give the compound **1** (0.4 mg) as a blue powder. Analytical HPLC: $t_R = 19.4$ min. ESI-TOF-MS *m/z* calcd for $C_{139}H_{169}N_{44}O_{27}S_2{}^{3+}$ $[M + 3H]^{3+}$ 983.4210 found 983.3976.

Synthesis of the Conjugate 2 (Deamino TH59TR). The compound **11** (0.6 mg, 2.7×10^{-4} mmol) and the powdered TR succinimidyl ester (0.3 mg, 3.7×10^{-4} mmol) was dissolved in DMF (200 μL) and DIEA (4 μL), followed by mixing for 4 h at room temperature with shielding the light. The reaction mixture was purified as described above to give the compound **2** (0.8 mg) as a blue powder. Analytical HPLC: $t_R = 20.7$ min. ESI-TOF-MS *m/z* calcd for $C_{139}H_{168}N_{43}O_{27}S_2{}^{3+}$ $[M + 3H]^{3+}$ 978.4173 found 978.3985.

Synthesis of the Conjugate 3 (TH59Cy3). First Cy3-NHS ester (4 mg) in DMF (80 μL) was prepared. The compound **10** (1.0 mg, 4.4×10^{-4} mmol) was dissolved in DMF (199.8 μL) and DIEA (0.2 μL), and the solution of Cy3-NHS ester (20.4 μL, 1.3×10^{-3} mmol) was added to that of **10**, followed by mixing for 4 h at room temperature with shielding the light. The reaction mixture was purified as described above to give the compound **3** (1.6 mg) as a red powder. Analytical HPLC: $t_R =$ 19.2 min. ESI-TOF-MS *m/z* calcd for $C_{132}H_{164}N_{43}O_{21}{}^{3+}$ $[M + 2H]^{3+}$ 895.7690 found 895.7559.

Synthesis of the Conjugate 4 (Deamino TH59Cy3). The compound **11** (0.9 mg, 4.0×10^{-4} mmol) was dissolved in DMF (180 μL) and DIEA (20 μL), and the solution of Cy3-NHS ester (40 μL, 2.6×10^{-3} mmol.) was added to that of **11**, followed by mixing for 5 h at room temperature with shielding the light. The reaction mixture was purified as described above to give the compound **4** (1.5 mg) as a red powder. Analytical HPLC: $t_R = 20.5$ min. ESI-TOF-MS *m/z* calcd for $C_{132}H_{163}N_{42}O_{21}{}^{3+}$ $[M + 2H]^{3+}$ 890.7654 found 890.7651.

2.4.4 Materials for Telomere Staining

Cell images were recorded with DeltaVision (Applied Precision). Mouse MC12 cells were gifts from Dr. Kimura, Osaka University. Hela1.3 cells were generous gifts of Dr. T. de Lange (Rockefeller University). DMEM medium was purchased from Life technology. Normal goat serum (NGS) was from Millipore. Nocodazole, paraphenylene diamine, and Triton X-100 were from Sigma. $NaBH_4$ was from Fluka. DAPI was from Roche. Rabbit anti-TRF1 antibody was a generous gift of Dr. F. Ishikawa (Kyoto University). Secondary antibody (goat anti-4rabbit Alexa 488) was from Molecular Probes. All other materials were from standard suppliers (highest quality available).

2.4.5 Telomeres Staining

Telomere Staining of Mouse MC12 Cell Spreads with Fluorescent Polyamides [24]. Mouse MC12 cells were maintained in DMEM medium containing 10% fetal bovine serum (FBS) at 37 °C (5% CO_2). Mouse MC12 cells were blocked mitotically by adding 0.1 μg/mL nocodazole for 1 h. The cells were swollen by treatment with a hypotonic buffer (0.075 M KCl) for 15 min at room temperature, fixed with a methanol/acetic acid (3:1) solution for 5 min. After centrifugation, cell pellet was again suspended in the methanol/acetic acid solution. The cell suspension was spread on coverslips, and air-dried for 30 min. The cell spread coverslips were kept at 4 °C until use.

The air-dried spread coverslips were soaked in TEN buffer (10 mM Tris-HCl, pH7.5, 1 mM EDTA and 100 mM NaCl) over night at 4 °C before use. For blocking, the spread coverslips were treated with 10% NGS in TEN buffer for 30 min at room temperature. After brief washing with TE buffer (10 mM Tris-HCl, pH7.5, 1 mM EDTA), the spread coverslips were incubated with 10% NGS, 500 nM polyamides and 0.5 μg/ml DAPI in TE buffer for 1 h at room temperature. After washing with TEN buffer (five times for 3 min), the spread coverslips were mounted in PPDI solution [10 mM HEPES pH 7.5, 100 mM KCl, 1 mM $MgCl_2$, 80% glycerol, 1 mg/mL paraphenylene diamine] and the coverslips were sealed with a nailporish. Sectioning images were recorded with DeltaVision and deconvoluted to eliminate 'out-of-focus' blur [25, 26]. The deconvoluted images were projected to obtain the maximum intensity ('Quick Projection' tool) of telomeric signals.

Co-localization of the Signal of Telomere-Specific Polyamides and TRF1. HeLa1.3 cells were maintained at 37 °C (5% CO_2) in DMEM containing 10% FBS. For polyamide staining, the HeLa cells were grown on coverslips coated with polylysine. The cell coverslips were washed in phosphate-buffered saline (PBS) and fixed with 2% formaldehyde in PBS for 15 min at room temperature. The fixed cell coverslips were then treated with 0.5 mg/ml $NaBH_4$ in PBS for 10 min, rinsed with PBS (twice for 5 min) and permeabilized with 0.5% Triton X-100 in PBS. After washing with PBS twice, and kept at 4 °C for storage. For fixation process, when we used XBE2 buffer (10 mM HEPES pH 7.5, 2 mM $MgCl_2$, 100 mM KCl, 5 mM EGTA) as described in ref. [12] and [26], instead of PBS, a similar telomere staining result was obtained (data not shown).

For blocking, the cell coverslips were treated with 10% NGS in TE buffer for 30 min at room temperature. After brief rinse with TE buffer, the cells were incubated with 10% NGS, rabbit anti-TRF1 antibody, 500 nM polyamide in DMF and 0.5 μg/ml

DAPI in TE buffer at 37 °C for 1 h. After washing with TEN200 (10 mM Tris-HCl, pH7.5, 1 mM EDTA and 200 mM NaCl) (five times for 3 min), the cells were incubated with secondary antibody (500-fold diluted goat anti-rabbit Alexa 488) and 1% NGS in TEN buffer for 1 h, and followed by five washes (3 min) with TEN200. The mounting and subsequent image acquisitions were performed as described above. Tight crosslinking by fresh formaldehyde solution tended to prevent the telomere staining. In such case, incubation of the fixed cells 37 °C over night facilitated the telomere staining (not shown).

For comparison between the **TH59TR** and TRF1-antibody signals, they were extracted from the Fig. 2.4a image as described previously [12]. The telomere signals yielded by the anti-TRF1 antibody and **TH59TR** were extracted based on threshold values using the Softworks software (Applied Precision). The maximum intensity values of Alexa 488 (anti-TRF1 antibody) and Texas Red signals in the extracted telomere regions were then used as telomere signals. Note that the mean values of Alexa 488 (anti-TRF1 antibody) and Texas Red signals in the extracted telomere regions also produced a similar correlation to Fig. 2.5 (not shown). Intensities of the corresponding **TH59TR** and TRF1-antibody signals were compared and shown as a scattered plot. Correlation coefficient was calculated using R software.

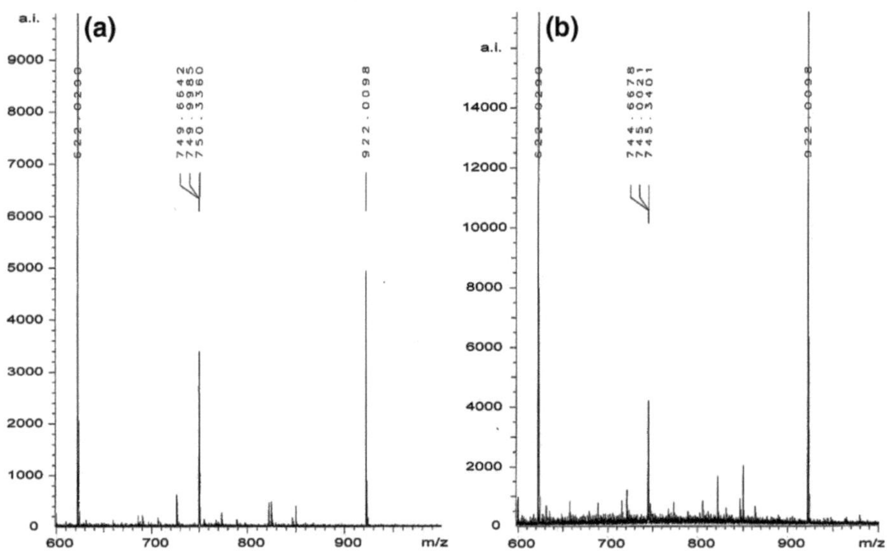

Fig. S2.1 ESI-TOF MS spectra of the compound **10** (**a**) and **11** (**b**). Copyright 2013 American Chemical Society

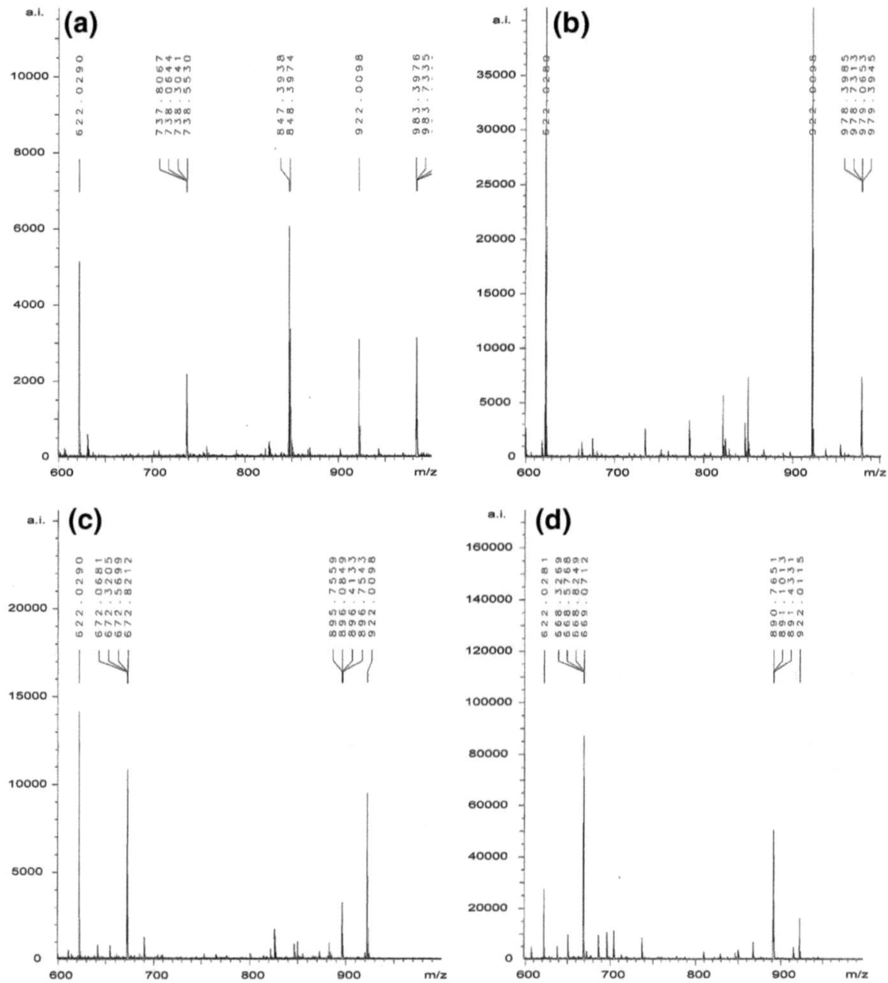

Fig. S2.2 ESI-TOF MS spectra of the conjugate **1–4** (a: **1**, b: **2**, c: **3**, d: **4**). Copyright 2013 American Chemical Society

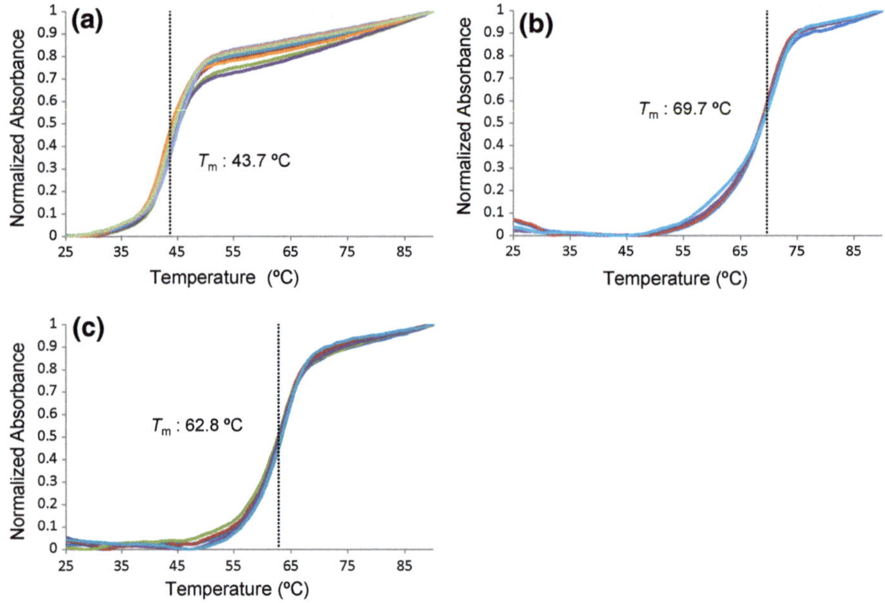

Fig. S2.3 UV melting profiles at 260 nm for ODN-3/4 (**a**), **TH59**-ODN-3/4 complex (**b**), **Deamino TH59**-ODN-3/4 (**c**). Copyright 2013 American Chemical Society

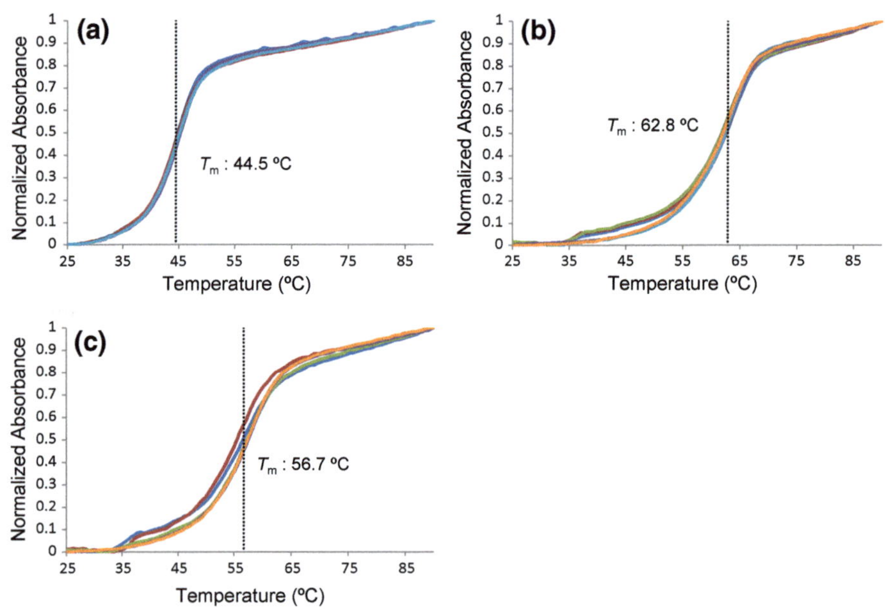

Fig. S2.4 UV melting profiles at 260 nm for ODN-3/4 (**a**), **TH59**-ODN-3/4 complex (**b**), **Deamino TH59**-ODN-3/4 (**c**). Copyright 2013 American Chemical Society

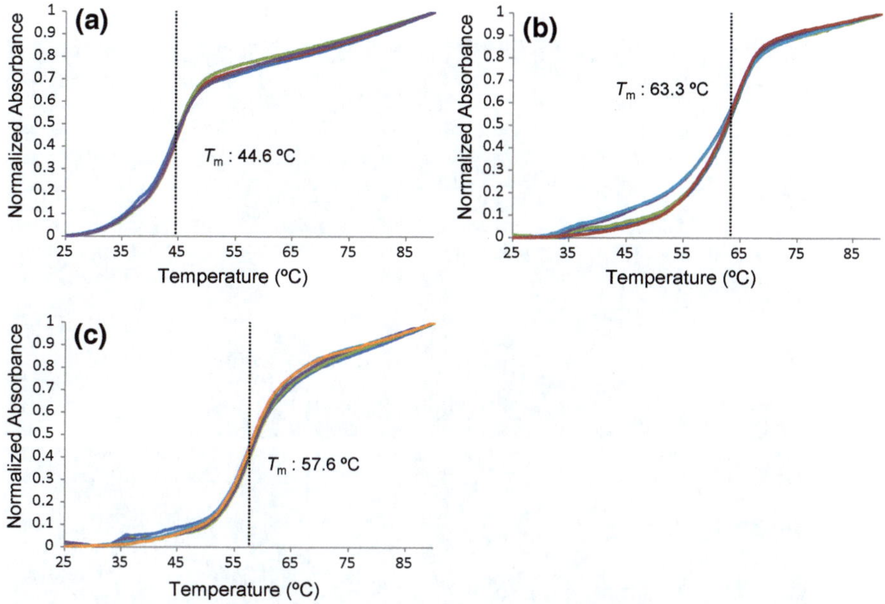

Fig. S2.5 UV melting profiles at 260 nm for ODN-5/6 (**a**), **TH59**-ODN-5/6 complex (**b**), **Deamino TH59**-ODN-5/6 (**c**). Copyright 2013 American Chemical Society

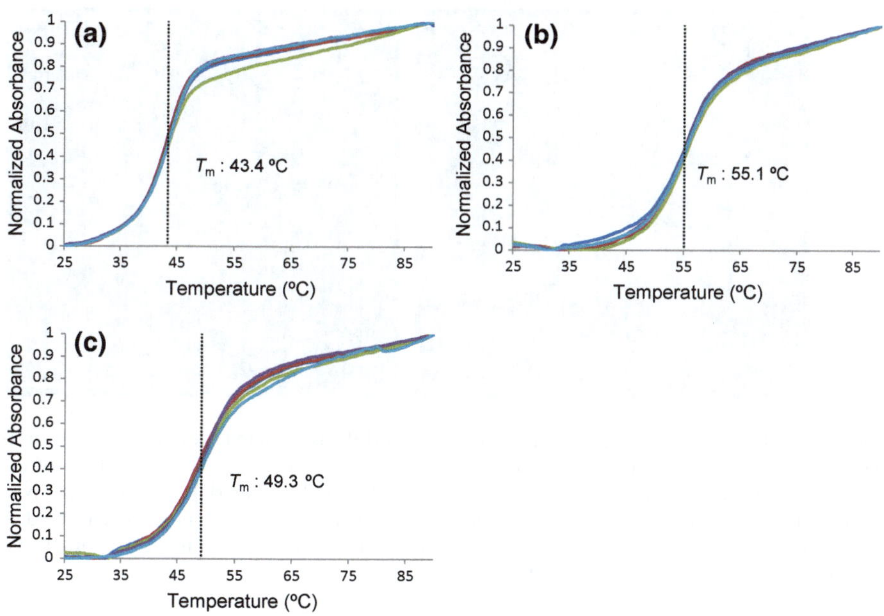

Fig. S2.6 UV melting profiles at 260 nm for ODN-7/8 (**a**), **TH59**-ODN-7/8 complex (**b**), **Deamino TH59**-ODN-7/8 (**c**). Copyright 2013 American Chemical Society

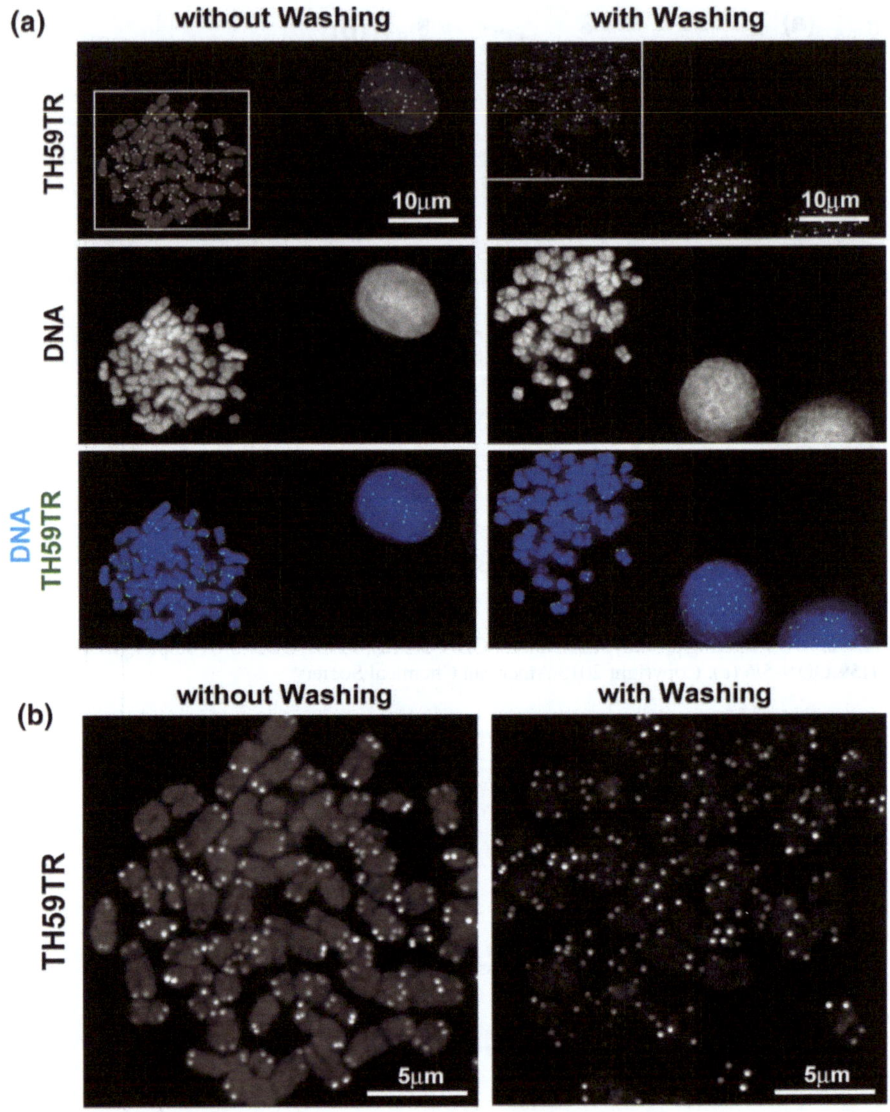

Fig. S2.7 Effect of washing on telomere staining by **TH59TR**. **a** Spreads of HeLa1.3 cells were stained with **TH59TR** with (right) or without (left) a washing process. The preparation of cell spreads and **TH59TR** staining were performed as described in the "Materials and Methods". Note that the telomeric signals were detected without the washing process, but the process helped reduce the background or nonspecific PI polyamide binding. **b** Enlarged images of the regions marked in Panel (**a**). Copyright 2013 American Chemical Society

References

1. Blackburn EH (2010) Angew Chem Int Ed 49:7405–7421
2. Nandakumar J, Cech TR (2013) Nat Rev Mol Cell Biol 14:69–82
3. Zakian VA (2012) Exp Cell Res 318:1456–1460
4. Smogorzewska A, de Lange T (2004) Annu Rev Biochem 73:177–208
5. Smogorzewska A, van Steensel B, Bianchi A, Oelmann S, Schaefer MR, Schnapp G, de Lange T (2000) Mol Cell Biol 20:1659–1668
6. (a) Dervan PB (2001) Bioorg Med Chem 9:2215–2235; (b) Dervan PB, Edelson BS (2003) Curr Opin Struc Biol 13:284–299; (c) Dervan PB, Doss RM, Marques MA (2005) Curr Med Chem -Anti-Cancer Agents 5:373–387; (d) Blackledge MS, Melander C (2013) Bioorg Med Chem 21:6101–6114
7. White S, Szewczyk JW, Turner JM, Baird EE, Dervan PB (1998) Nature 391:468–471
8. (a) Turner JM, Swalley SE, Baird EE, Dervan PB (1998) J Am Chem Soc 120:6219–6226; (b) Wang CC, Ellervik U, Dervan PB (2001) Bioorg Med Chem 9:653–657
9. (a) de Clairac RPL, Geierstanger BH, Mrksich M, Dervan PB, Wemmer DE (1997) J Am Chem Soc 119:7909–7916; (b) Murty MSRC, Sugiyama H (2004) Biol Pharm Bull 27:468–474
10. (a) Herman DM, Turner JM, Baird EE, Dervan PB (1999) J Am Chem Soc 121:1121–1129; (b) Melander C, Herman DM, Dervan PB (2000) Chem Eur J 6:4487–4497; (c) Morinaga H, Bando T, Takagaki T, Yamamoto H, Hashiya K, Sugiyama H (2011) J Am Chem Soc 133:18924–18930; (d) Li BC, Montgomery DC, Puckett JW, Dervan PB (2013) J Org Chem 78:124–133
11. (a) Herman DM, Baird EE, Dervan PB (1999) Chem Eur J 5:975–983; (b) Kers I, Dervan PB (2002) Bioorg Med Chem 10:3339–3349; (c) Schaal TD, Mallet WG, McMinn DL, Nguyen NV, Sopko MM, John S, Parekh BS (2003) Nucleic Acids Res 31:1282–1291; (d) Sasaki S, Bando T, Minoshima M, Shinohara K, Sugiyama H (2008) Chem Eur J 14:864–870
12. Maeshima K, Janssen S, Laemmli UK (2001) EMBO J 20:3218–3228
13. (a) Bando T, Narita A, Saito I, Sugiyama H (2003) J Am Chem Soc 125:3471–3485; (b) Bando T, Sugiyama H (2006) Acc Chem Res 39:935–944; (c) Bando T, Sasaki S, Minoshima M, Dohno C, Shinohara K, Narita A, Sugiyama H (2006) Bioconjugate Chem 17:715–720. (d) Takagaki T, Bando T, Sugiyama H (2012) J Am Chem Soc 134:13074–13081
14. (a) Vaijayanthi T, Bando T, Pandian GN, Sugiyama H (2012) ChemBioChem 13:2170–2185; (b) Vaijayanthi T, Bando T, Hashiya K, Pandian GN, Sugiyama H (2013) Bioorg Med Chem 21:852–855; (c) Su W, Bagshaw CR, Burley GA (2013) Sci Rep 3:1883
15. (a) Ohtsuki A, Kimura MT, Minoshima M, Suzuki T, Ikeda M, Bando T, Nagase H, Shinohara K, Sugiyama H (2009) Tetrahedron Lett 50:7288–7292; (b) Pandian GN, Shinohara K, Ohtsuki A, Nakano Y, Minoshima M, Bando T, Nagase H, Yamada Y, Watanabe A, Terada N, Sato S, Morinaga H, Sugiyama H (2011) ChemBioChem 12:2822–2828. (c) Pandian GN, Ohtsuki A, Bando T, Sato S, Hashiya K, Sugiyama H (2012) Bioorg Med Chem 20:2656–2660; (d) Pandian GN, Nakano Y, Sato S, Morinaga H, Bando T, Nagase H, Sugiyama H (2012) Sci Rep 2:544
16. Fujimoto J, Bando T, Minoshima M, Kashiwazaki G, Nishijima S, Shinohara K, Sugiyama H (2008) Bioorg Med Chem 16:9741–9744
17. (a) Takahashi R, Bando T, Sugiyama H (2003) Bioorg Med Chem 11:2503–2509; (b) Sasaki S, Bando T, Minoshima M, Shimizu T, Shinohara K, Takaoka T, Sugiyama H (2006) J Am Chem Soc 128, 12162–12168; (c) Kashiwazaki G, Bando T, Shinohara K, Minoshima M, Nishijima, S, Sugiyama S (2009) Bioorg Med Chem 17:1393–1397; (d) Kashiwazaki G, Bando T, Shinohara K, Minoshima M, Kumamoto H, Nishijima S, Sugiyama H (2010) Bioorg Med Chem 18:2887–2893
18. Wurtz NR, Turner JM, Baird EE, Dervan PB (2001) Org Lett 3:1201–1203
19. (a) Rucker VC, Foister S, Melander C, Dervan PB (2003) J Am Chem Soc 125:1195–1202. (b) Rucker VC, Dunn AR, Sharma S, Dervan PB, Gray HB (2004) J Phys Chem B 108:7490–7494

20. (a) Pilch DS, Poklar N, Gelfand CA, Law SM, Breslauer KJ, Baird EE, Dervan PB (1996) Proc Natl Acad Sci USA 93:8306–8311; (b) Muzikar KA, Meier JL, Gubler DA, Raskatov JA, Dervan PB (2011) Org Lett 13:5612–5615
21. Kumar S, Xue L, Arya DP (2011) J Am Chem Soc 133:7361–7375
22. (a) Herman DM, Baird EE, Dervan PB (1998) J Am Chem Soc 120:1382–1391; (b) Dose C, Farkas ME, Chenoweth DM, Dervan PB (2008) J Am Chem Soc 130:6859–6866
23. Aubert G, Hills M, Lansdorp PM (2012) Mutat Res 730:59–67
24. Yoshida I (2002) Cytogenet. Genome Res 99:44–51
25. Maeshima K, Laemmli UK (2003) Dev Cell 4:467–480
26. Maeshima K, Yahata K, Sasaki Y, Nakatomi R, Tachibana T, Hashikawa T, Imamoto F, Imamoto N (2006) J Cell Sci 119:4442–4451

Chapter 3
Tandem Trimer Pyrrole-Imidazole Polyamide Probes Targeting 18 Base Pairs in Human Telomere Sequences

Abstract The binding of molecules to specific DNA sequences is important for imaging genome DNA and for studying gene expression. Increasing the number of base pairs targeted by these molecules would provide greater specificity. *N*-methylpyrrole–*N*-methylimidazole (Py–Im) polyamides are one type of such molecules and can bind to the minor groove of DNA in a sequence-specific manner without causing denaturation of DNA. Our recent work has demonstrated that tandem hairpin Py–Im polyamides conjugated with a fluorescent dye can be synthesized easily and can serve as new probes for studying human telomeres under mild conditions. Herein, to improve their selectivities to telomeres by targeting longer sequences, we designed and synthesized a fluorescent tandem trimer Py–Im polyamide probe, comprising three hairpins and two connecting regions (hinges). The new motif bound to 18 bp dsDNA in human telomeric repeats $(TTAGGG)_n$, the longest sequence for specific binding reported for Py–Im polyamides. We compared the binding affinities and the abilities to discriminate mismatch, the UV–visible absorption and fluorescence spectra, and telomere staining in human cells between the tandem trimer and a previously developed tandem hairpin. We found that the tandem trimer Py–Im polyamide probe has higher ability to recognize telomeric repeats and stains telomeres in chemically fixed cells with lower background signal.

Keywords Pyrrole–imidazole polyamides · Tandem trimer polyamides · Solid-phase peptide synthesis · Telomere · Cellular imaging

3.1 Introduction

Telomeres are localized at the ends of chromosomes and play important roles in the stability and replication of chromosomes [1]. Human telomere DNA, whose sequences are the tandem repeats of 5′-TTAGGG-3′, has a duplex region and a 3′

This chapter is reprinted with permission from "Tandem Trimer Pyrrole-Imidazole Polyamide Probes Targeting 18 Base Pairs in Human Telomere Sequences" Kawamoto, Y.; Sasaki, A.; Hashiya, K.; Ide, S; Bando, T.; Maeshima, K; Sugiyama, H. *Chem. Sci.* **2015**, *6*, 2307–2312. Copyright 2015 The Royal Society of Chemistry.

Y. Kawamoto, *Synthesis and Biological Evaluation of Pyrrole–Imidazole Polyamide Probes for Visualization of Telomeres*, Springer Theses,
https://doi.org/10.1007/978-981-13-6912-4_3

overhang, which tends to form G-quadruplexes or t-loops [2]. In normal mammalian cells, the number of telomere repeats decreases with cell division, and this decrease is related to the aging process and cancer [3, 4]. Therefore, the telomere length is one important biomarker used in various diagnoses [2–5]. Fluorescent probes targeting telomeres have been developed to visualize and measure the telomere length, which is regulated by the shelterin complex [6].

Our group has focused on *N*-methylpyrrole (Py)–*N*-methylimidazole (Im) (Py–Im) polyamides, as reported by Dervan and co-workers. Py–Im polyamides bind strongly and sequence-specifically to the minor groove of dsDNA by recognizing Watson–Crick base pairs without causing DNA denaturation [7]. The antiparallel Im/Py pair recognizes a G·C base pair, whereas a Py/Py pair recognizes an A·T or T·A base pair [7a, b]. C-terminal β-alanine and the γ-turn moiety comprising γ-aminobutyric acid or 2, 4-diaminobutyric acid (Dab) also recognize A·T or T·A base pairs, and (R)-Dab affords a higher binding affinity comparable to that of γ-aminobutyric acid [8].

Py–Im polyamides can be synthesized by machine-assisted Fmoc solid-phase peptide synthesis (SPPS) [9]. Various structures of Py–Im polyamides, such as linear [10], hairpin [11], and cyclic [12], have been synthesized to provide efficient binding. In particular, hairpin Py–Im polyamides are versatile and can be used as sequence-specific functional conjugates [13]. To gain specificity, Py–Im polyamides targeting longer sequences (≥10 bp) have been developed [10, 14–17]. DNA ligands recognizing ≥16 bp are required to target a single site in the human genome, which comprises about 3 billion base pairs. Dervan and co-workers reported a linear Py–Im polyamide that formed a homodimer with two molecules and then bound to 16 bp sequences in the regulatory region of the HIV-1 genome [10b]. Our laboratory has reported a cysteine-derived Py–Im polyamide dimer that binds to 16 bp with one molecule [14c].

Tandem hairpin Py–Im polyamides targeting ≥10 bp, which comprise two hairpin moieties and a hinge segment between two hairpins, have been developed [15, 16]. For instance, Maeshima, Janssen, and Laemmli described tandem hairpin Py–Im polyamide probes that could bind to 12 bp sequences in the duplex region of the human telomere repetitive sequence TTAGGG and could stain telomeres in chemically fixed cells [16]. Solid-phase synthesis requires many steps, which leads to a lower yield, and there are fewer reports on the targeting longer sequences with Py–Im polyamides than those described for the binding of conventional polyamides to shorter sequences.

Recently, our group proposed a new method for the synthesis of tandem hairpin Py–Im polyamide **TH59** (Fig. 3.1) using a new building block. The block comprises the sequence from Dab to the *N*-terminus of a hairpin moiety, and is used as a unit of SPPS to decrease the procedure on solid support. Using this methodology, we synthesized fluorescent telomere probes [17a, b] and alkylation agents [17c, d]. Human telomeres bound by the probes could be costained with anti-TRF1 antibody under mild conditions without breaking the structure of the telomeres, showing that the telomere length was related to the abundance of the shelterin complex [17a]. This point is an advantage over the PNA FISH method, which requires denaturation under harsh conditions [6b]. More recently, the fluorescent group and the hinge region were optimized to target telomeres in cells more selectively [17b].

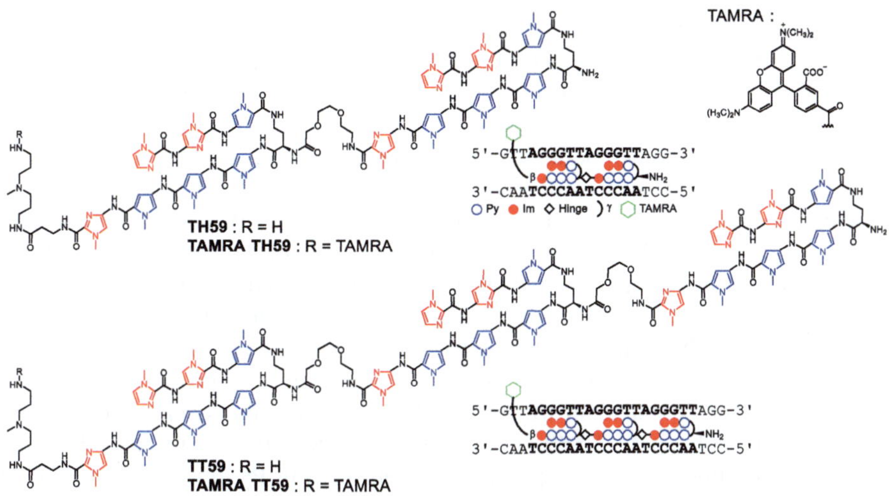

Fig. 3.1 Chemical structures of Py–Im polyamides **TH59**, **TT59**, **TAMRA TH59** and **TAMRA TT59** targeting the human telomere sequence and their ball-and-stick representations. Copyright 2015 The Royal Society of Chemistry

Herein, as one of the optimizations of the Py-Im polyamide probes, we describe our synthesis of the new type of Py–Im polyamides, defined as tandem trimers, whose components are three hairpin subunits and two hinges. Our aim was to increase their selectivities to telomeres. After the optimization of SPPS, we synthesized the new tandem trimer Py–Im polyamide **TT59**, which targeted 18 bp in the human telomere repetitive sequence TTAGGG. To our knowledge, 18 bp is the longest specific binding sequence for Py–Im polyamides. Surface plasmon resonance (SPR) analysis was used to assess the binding affinity and specificity of new Py–Im polyamide **TT59** toward telomeric repeat sequences. After conjugation of **TH59** and **TT59** with fluorescent 5-carboxytetramethylrhodamine (TAMRA) [17b], we obtained fluorescent telomere probes **TAMRA TH59** and **TAMRA TT59**, respectively (Fig. 3.1). We found that the new tandem trimer probe **TAMRA TT59** stained telomeres in chemically fixed cells and produced a lower background because of the reduction in nonspecific binding compared with the conventional probe **TAMRA TH59**. These results suggest that the new longer Py–Im polyamides recognize the target sequences in cells with greater specificity.

3.2 Results and Discussion

3.2.1 Synthesis of Py–Im Polyamides

Based on the previous methodology using the synthesized building block **1** corresponding to (*R*)-2,4-diaminobutyric acid turn and three-ring cycle polyamide at the side of N-terminal as one of the units [17a], synthesis of fluorescent Py–Im polyamide probes was performed after optimization (Scheme 3.1). These syntheses have suffered from low yield, presumably due to the contamination of moisture. Therefore, the solid-phase peptide synthesizer was placed in a vinyl box with dehumidifiers. Machine-assited Fmoc SPPS began from Fmoc-β-Ala-Wang resin. Each Fmoc group was deprotected 20% piperidine/1-methyl-2-pyrrolidone (NMP) and then the deprotected amino group reacted with the next Fmoc unit (Fmoc-Py-OH, Fmoc-PyIm-OH, Fmoc-mini-PEG-OH or **1**) activated with *N*, *N*-diisopropylethylamine (DIEA) and HCTU in NMP. Procedures (i)-(xix) were performed for the synthesis of tandem dimer of hairpins on solid support. Though conventional **TH59** could be obtained through the cleavage of the dimer from resin with 3, 3′-diamino-*N*-methyldipropylamine, one repeat of procedures (x)–(xix) was required for new tandem trimer **TT59**. As a result, the synthesis of tandem trimer Py–Im polyamide **TT59** succeeded for the first time with an 8.5% yield for SPPS, and the yield of the previous tandem hairpin Py–Im polyamide **TH59** increased from 6.9 to 29% [17a]. The obtained Py–Im polyamides **TH59** and **TT59** were conjugated with 5-carboxytetramethylrhodamine succinimidyl ester in *N*, *N*-dimethylformamide (DMF) and DIEA to afford the fluorescent Py–Im polyamide probes **TAMRA TH59** and **TAMRA TT59**, respectively. Protection of the α-amino group of Dab was not required because of steric hindrance. The ESI-TOF MS spectra and analytical HPLC profiles of **TH59**, **TT59**, **TAMRA TH59** and **TAMRA TT59** are shown in Figs. S3.1 and S3.2 respectively.

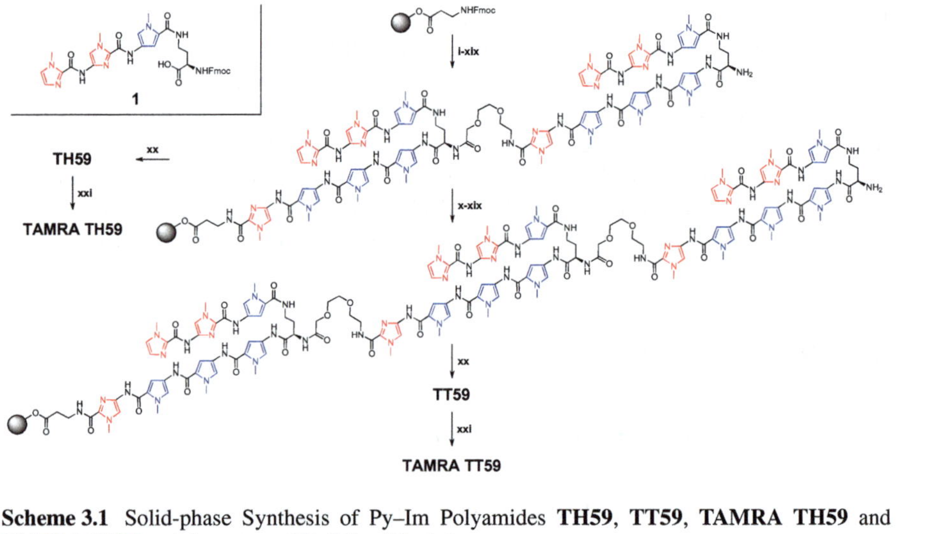

Scheme 3.1 Solid-phase Synthesis of Py–Im Polyamides **TH59**, **TT59**, **TAMRA TH59** and **TAMRA TT59** and Structure of Building Block **1**

Reagents and conditions: (i) 20% piperidine, NMP; (ii) Fmoc-PyIm-CO_2H, HCTU, DIEA, NMP; (iii) 20% piperidine, NMP; (iv) Fmoc-Py-CO_2H, HCTU, DIEA, NMP; (v) 20% piperidine, NMP; (vi) Fmoc-Py-CO_2H, HCTU, DIEA, NMP; (vii) 20% piperidine, NMP; (viii) **1**, HCTU, DIEA, NMP; (ix) 20% piperidine, NMP; (x) Fmoc-mini-PEG-OH, HCTU, DIEA, NMP; (xi) 20% piperidine, NMP; (xii) Fmoc-PyIm-CO_2H, HCTU, DIEA, NMP; (xiii) 20% piperidine, NMP; (xiv) Fmoc-Py-CO_2H, HCTU, DIEA, NMP; (xv) 20% piperidine, NMP; (xvi) Fmoc-Py-CO_2H, HCTU, DIEA, NMP; (xvii) 20% piperidine, NMP; (xviii) **1**, HCTU, DIEA, NMP; (xix) 20% piperidine, NMP; (xx) 3,3′-diamino-*N*-methyldipropylamine, 55 °C; (xxi) 5-carboxytetramethylrhodamine, succinimidyl ester, DIEA, DMF.

3.2.2 Optical Characteristics of the Fluorescent Probes

The UV–visible absorption spectra of probes **TAMRA TH59** and **TAMRA TT59** are shown in Fig. S3.3a. The absorption maximum at ~555 nm was due to the fluorescent group TAMRA. The absorption maximum at ~310 nm was derived from the *N*-methylpyrrole and *N*-methylimidazole moieties.

Figure S3.3b shows the fluorescence spectra of probes **TAMRA TH59** and **TAMRA TT59** in the absence and presence of 1.0 or 2.0 equivalent oligonucleotide (ODN)-1/2, whose sequences are shown in Table S3.1. Addition of ODN-1/2 to both probes provided little change in the fluorescence intensity and quench of TAMRA was not observed in the absence of ODN-1/2. In contrast, Dervan, Gray and co-workers showed that TAMRA connected to *N*-methylpyrrole in the polyamide backbone was quenched because of the electron transfer from Py to TAMRA in the absence of dsDNA [18]. It is presumed that the distance between TAMRA attaching to C-terminal through 3, 3′-diamino-*N*-methyldipropylamine linker and polyamide was longer and therefore the electron transfer triggering quench became difficult to occur.

3.2.3 Binding Affinities to Match Sequence and Discrimination of Mismatch Sequences

Binding affinities and the abilities to discriminate mismatch sequences of **TH59** and **TT59** were evaluated with SPR experiments [17b, 19]. Three kinds of 5′-biotinylated hairpin DNAs ODN-3, 4 and 5 containing the binding site of **TT59** were prepared by the immobization to sensor chips through a biotin-avidin system. The sequences are shown in Tables 3.1 and S3.1. ODN-3, 4 and 5 had no mismatch, 1 bp mismatch at the C-terminal of **TT59** and 1 bp mismatch at the center of **TT59**, respectively. Tandem hairpin **TH59** and trimer **TT59** were passed to DNAs on sensor chips and resulting sensorgrams (shown in Figs. S3.4 and S3.5) and values (summarized in

Table 3.1) were obtained. Comparing the values of **TT59** in the matching sequence with those of conventional tandem hairpin **TH59**, the association of **TT59** was slower and thus its binding affinity was weaker. One considerable reason was that bigger structure of tandem trimer **TT59** decreased its own accessibility to the matching sequence. However, since the K_D value of **TT59** and ODN-3 was 3.6×10^{-9} M, the tandem trimer had high binding affinity comparable with other Py-Im polyamides [7]. To evaluate the discriminations, specificities were calculated by dividing the K_D value of the mismatch DNA (ODN-4 or 5) by that of the match DNA (ODN-3). The specificity value of tandem trimer **TT59** was bigger than that of **TH59**, suggesting that extending the targeted base pairs could increase the sequence specificities.

3.2.4 Human Telomere Staining with Two Py–Im Polyamide Probes

We doubly stained human HeLa 1.3 cell spreads, which are used for clinical karyotype tests, with 4′, 6-diamidino-2-phenylindole (DAPI) and Py–Im polyamide probes **TAMRA TH59** and **TAMRA TT59** to compare their abilities to stain telomeres selectively. Cell spreads were prepared by methanol/acetic acid fixation. To ensure a proper comparison between **TAMRA TH59** and **TAMRA TT59**, the images without deconvolution, differently from our previous study [17a], were presented in Fig. 3.2. Furthermore, to demonstrate reproducibility of the result, additional image data was depicted in Fig. S3.6. DAPI visualized chromosomal regions and nuclei, and Py–Im polyamide probes showed intense foci. Two foci were observed at every chromosomal end, indicating that both probes could stain telomeres selectively. However, the signal intensities appeared to be greater for the probe **TAMRA TH59** [17b] than for **TAMRA TT59**, presumably because more **TAMRA TH59** molecules bound to telomeres compared with **TAMRA TT59**. On the other hand, as shown in Fig. 3.2b, substantial polyamide signals were observed on the chromosome body when treating cells with **TAMRA TH59**. By contrast, lower-intensity signals on the chromosome body region were observed in the case of **TAMRA TT59**. Background polyamide signals except intense foci derived from nonspecific binding. Cell spreads images indicated **TAMRA TT59** had higher selectivity to telomeres, suggesting that non-specific binding was reduced by increasing the number of base pairs bound by Py–Im polyamides.

This result was supported by telomere staining images of HeLa 1.3 cells fixed in formaldehyde (Figs. 3.3 and S3.7), a commonly used fixation method in cell biology. Many sharp polyamide signals were detected by staining with both **TAMRA TH59** and **TAMRA TT59** in nuclei. By contrast, throughout the nucleus, **TAMRA TT59** signals were less intense and rarely merged with DAPI signals. For quantification of the telomere signals, they were extracted from Fig. 3.3 non-deconvolved images and the results are shown in Fig. 3.4 as surface plots and signal-to-noise ratio. The sharper surface plots and about 3.7-fold signal-to-noise ratio compared to **TAMRA TH59**

Table 3.1 Binding affinities of polyamides **TH59** and **TT59** against match sequence (ODN-3) or 1 bp mismatch sequence (ODN-4 or 5) [a]

	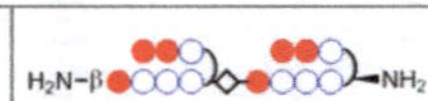 **TH59**					 **TT59**				
	k_a ($M^{-1}s^{-1}$)	k_d(s^{-1})	K_D(M)	χ^2	Specificity	k_a ($M^{-1}s^{-1}$)	k_d (s^{-1})	K_D(M)	χ^2	Specificity
5'-biotin-GGTTAGGGTTAGGGTTAGGGTTAGG TT 3'-CCAATCCCAATCCCAATCCCAATCC ODN-3	2.0 × 10^5	3.3 × 10^{-4}	1.7 × 10^{-9}	0.69	–	9.8 × 10^4	3.5 × 10^{-4}	3.6 × 10^{-9}	0.53	–
5'-biotin-GGTTAGAGTTAGGGTTAGGGTTAGG TT 3'-CCAATCTCAATCCCAATCCCAATCC ODN-4	3.4 × 10^5	6.5 × 10^{-4}	1.9 × 10^{-9}	0.36	1.1	5.3 × 10^4	4.9 × 10^{-4}	9.3 × 10^{-9}	0.60	2.6
5'-biotin-GGTTAGGGTTAGAGTTAGGGTTAGG TT 3'-CCAATCCCAATCTCAATCCCAATCC ODN-5	1.3 × 10^5	3.7 × 10^{-4}	2.8 × 10^{-9}	0.20	1.6	4.7 × 10^4	4.2 × 10^{-4}	9.0 × 10^{-9}	0.33	2.5

[a] Specificity was calculated by K_D(Mismatch)/K_D(Match).

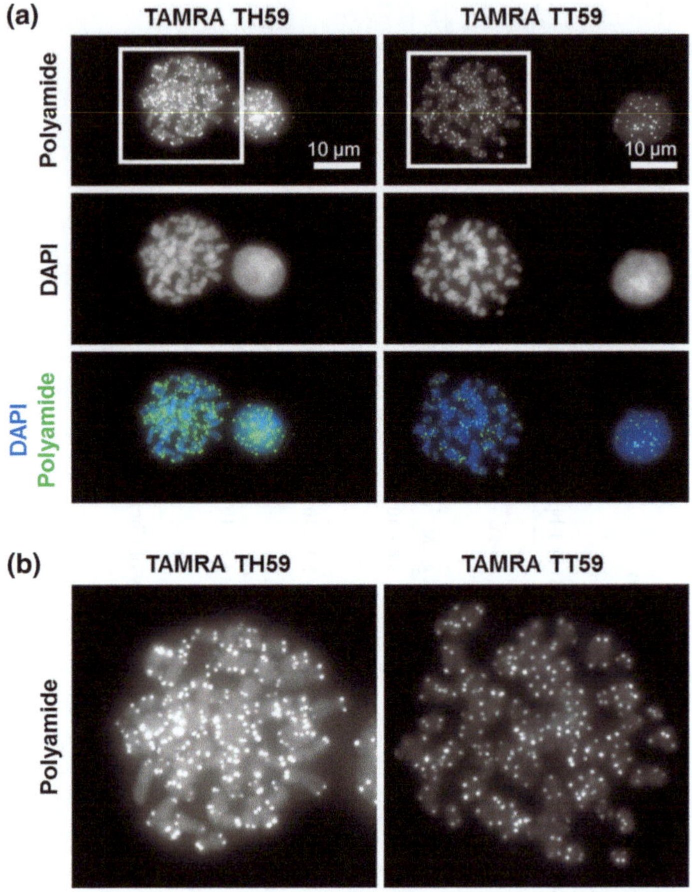

Fig. 3.2 Telomere staining of HeLa 1.3 cell spreads with fluorescent polyamides. **a** The HeLa 1.3 cell spread was stained with the fluorescent polyamide (first row) and DAPI (second row). The merged images are shown in the third row. The first and second columns show the results for **TAMRA TH59** and **TAMRA TT59**, respectively. Enlarged images of the boxed regions in panel (**a**) are shown in panel (**b**). Copyright 2015 The Royal Society of Chemistry

also showed lower-intensity background signals of **TAMRA TT59**, suggesting its higher selectivity to telomeres. These images of cell staining were supported by the results of SPR analyses.

3.3 Conclusion

In this study, we have shown the new motif of Py–Im polyamide, which we call a tandem trimer, which comprises three hairpins and two hinge segments and targets

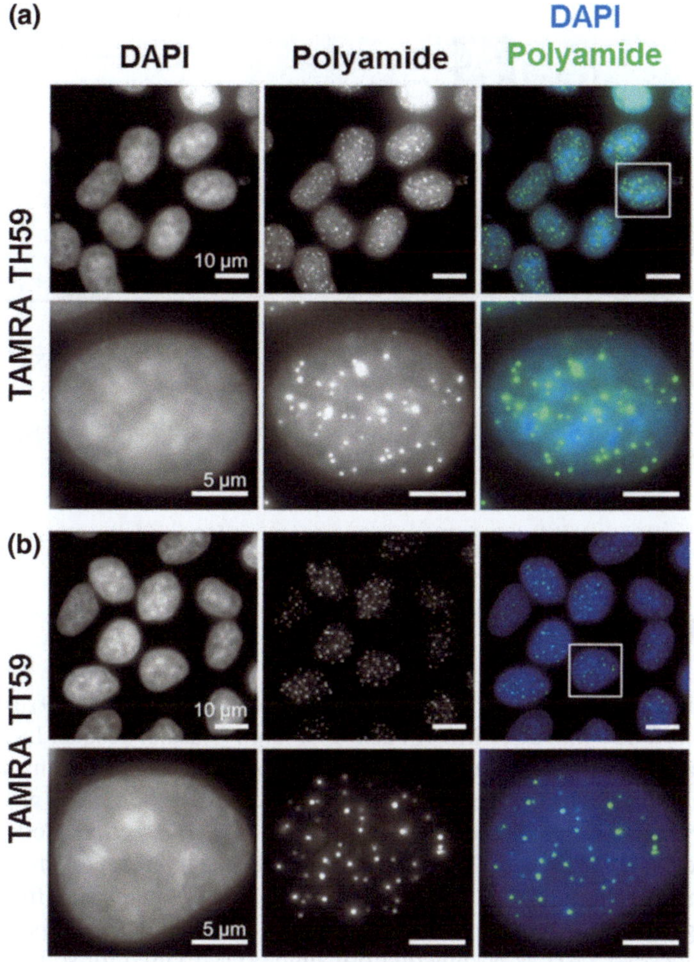

Fig. 3.3 Telomere staining of HeLa 1.3 cells with fluorescent polyamides. HeLa 1.3 cells were stained with DAPI (first column) and fluorescent polyamides (second column). The merged images are shown in the third column. Enlarged images of the boxed regions in the first row are shown in the second row. Cell images using **TAMRA TH59** and **TAMRA TT59** are shown in panels (**a**) and (**b**), respectively. Copyright 2015 The Royal Society of Chemistry

18 bp, the longest reported sequence for Py–Im polyamides. SPR analysis indicated that the tandem trimer Py–Im polyamide bound to 18 bp in human telomere repeats more specifically than did the previous tandem hairpin Py–Im polyamides. Moreover, the TAMRA-conjugated tandem trimer Py–Im polyamide probe highlighted the telomere foci clearly with lower intensity of background signals, showing its higher selectivity to telomeres. This new motif may be useful for highlighting specific regions clearly in human cells.

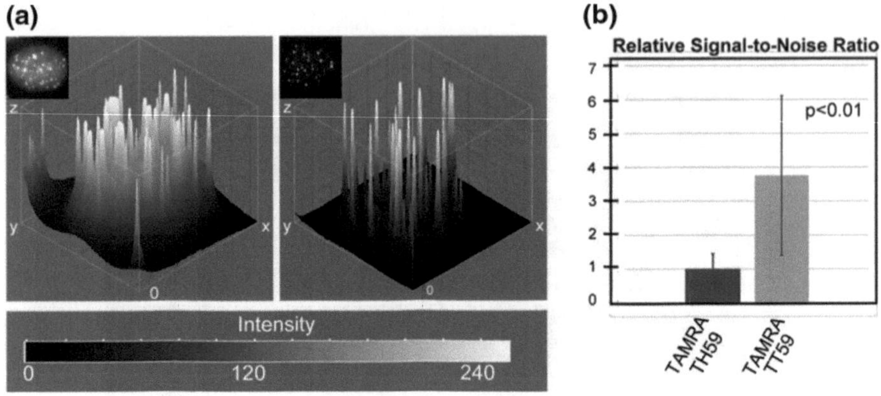

Fig. 3.4 **a** Surface plots based on the boxed images shown in Fig. 3.3. **b** Relative signal-to-noise ratios of the images. Copyright 2015 The Royal Society of Chemistry

3.4 Supporting Information

3.4.1 Materials

HPLC analysis was performed on a PU-2089 plus series system (JASCO) using COSMOSIL 4.6 × 150 mm $5C_{18}$-MS-II Packed Column (Nacalai Tesque, Inc.) in 0.1% TFA in water with acetonitrile as the eluent at a flow rate of 1.0 mL/min and a linear gradient elution of 0 to 50% acetonitrile in 20 min with detection at 254 nm. Purifications of Py–Im polyamides are mentioned in each synthetic section. Collected fractions were analyzed by ESI-TOF-MS (Bruker). For the collect mass measurement, external mass calibration (m/z: 622.0290 and 922.0098), purchased from Agilent Technologies, was used. Fmoc-mini-PEG-OH was purchased from Peptide International. Fmoc-Py-CO_2H, *N*,*N*-dimethylformamide (DMF) used for solid-phase synthesis, 1-methyl-2-pyrrolidone (NMP), trifluoroacetic acid (TFA), and piperidine were from Wako. HCTU was from Peptide Institute, Inc. Fmoc-β-Ala-Wang resin (0.60 mmol/g, 100–200 mesh) was from Novabiochem. *N*, *N*-diisopropylethylamine (DIEA) was from Nacalai Tesque, Inc. DMF used for solution-phase synthesis and for telomere-staining and acetonitrile were from Kanto Chemical Co., Inc. 3, 3′-diamino-*N*-methyldipropylamine was from Tokyo Chemical Industry Co., LTD. 5-carboxytetramethylrhodamine succinimidyl ester was from Molecular Probes. The Fmoc units **1** [17a] and Fmoc-PyIm-CO_2H [20] were synthesized using the previous method. SPPS was performed on a PSSM-8 (Shimadzu). Dehumidifiers were from Panasonic. Concentrations of Py–Im Polyamides were calculated with a Nanodrop ND-1000 spectrophotometer (Thermo Fisher Scientific Inc.) using an extinction coefficient of 9900 $M^{-1}cm^{-1}$ per one pyrrole or imidazole moiety at λ_{max} near 310 nm. Measurement of absorption spectra was performed on a Spectophotometer V-650 (JASCO) equipped with a thermocontrolled PAC-743R cell changer (JASCO) and a refrigerated and heating circulator F25-ED (Julabo). Fluorescence spectra were measured on a Spectrofluorometer FP-6300 (JASCO). The SPR assays were performed using a BIACORE X instrument (GE Healthcare) with 5′-biotinylated DNA oligonucleotides (JBioS), streptavidin-coated sensor chip SA (GE Healthcare) and HBS-EP buffer (GE Healthcare). All unmodified DNA oligonucleotides used in this study were purchased from Sigma. Cell images were recorded with DeltaVision (Applied Precision). HeLa 1.3 cells were generous gifts of Dr. T. de Lange (Rockefeller University). DMEM medium was purchased from Invitrogen. Normal goat serum (NGS) was from Millipore. Nocodazole, paraphenylene diamine, and Triton X-100 were from Sigma. $NaBH_4$ was from Fluka. DAPI was from Roche. All other solvents and materials were from standard suppliers (highest quality available).

3.4.2 Synthesis of Py–Im Polyamides

General Procedures of Fmoc Solid-Phase Peptide Synthesis. Each synthesis of Py–Im polyamides was performed with a computer-assisted operation system placed in a vinyl box with dehumidifiers by using Fmoc chemistry under low humidity (<20%). In the absence of dehumidifiers, the humidity was higher than 40% and the yield of **TH59** was only 6.9% [17a]. Furthermore the synthesis of **TT59** was impossible. An Fmoc unit in each step was set up to solve by NMP (1 mL) on the synthetic line. Fmoc units in each step were as follows: Fmoc-Py-CO_2H (77 mg, 0.21 mmol), Fmoc-PyIm-CO_2H (70 mg, 0.14 mmol), Fmoc-D-Dab(ImImPy)-OH (**1**) (70 mg, 0.10 mmol), and Fmoc-mini-PEG-OH (81 mg, 0.21 mmol). Before SPPS, Fmoc units were dissolved in NMP (1.0 mL) with HCTU and the amount of HCTU for Fmoc-Py-CO_2H, Fmoc-PyIm-CO_2H, **1**, and Fmoc-mini-PEG-OH was 88 mg (0.21 mmol), 60 mg (0.15 mmol), 40 mg (0.097 mmol), and 88 mg (0.21 mmol), respectively.

Procedures were as follows: twice deblocking for 4 min with 20% piperidine/NMP (0.50 mL), addition of 10% DIEA/NMP (400 μL, DIEA was 0.23 mmol) to Fmoc unit-HCTU mixture for activation, coupling for 60 min. Five times washing with DMF were performed during each step. After the last coupling, the deblocking of amino group at γ-turn was performed. All coupling were carried out with a single-coupling cycle. All lines were purged with solution transfers and bubbled by nitrogen gas for stirring the resin. After solid-phase synthesis, the resin was washed with methanol and dried *in vacuo*, followed by the cleavage of the compound from Wang resin using 0.40 mL 3, 3′-diamino-*N*-methyldipropylamine (96% assay) for 3 h at 55 °C. The resin was removed by filtration and washed thoroughly with dichloromethane, and the filtrate was concentrated *in vacuo*. The residue was dissolved in 0.50 mL dichloromethane and then 5.0 mL diethyl ether was added to the solution, followed by sonication for a few minutes to obtain the yellow residue. After discarding supernatant, the residue was dissolved in 0.50–1.0 mL dichloromethane-methanol mixture and then more than 10-fold volume of diethyl ether was added, followed by the centrifuge and removal of the supernatant. This process was repeated a few times until white precipitation was obtained. The dried crude precipitation was purified by reversed-phase chromatography.

Synthesis of Tandem Hairpin Py–Im Polyamide TH59. Solid-phase synthesis was performed with 43.0 mg Fmoc-β-Wang resin (0.60 mmol/g). The resulting crude polyamide (33.7 mg) was purified by reversed-phase flash chromatography. Purification was performed on CombiFlash Rf using a 4.3 g C18 RediSep Rf reversed-phase flash column (Teledyne Isco, Inc.) in 0.1% TFA in water with acetonitrile as the eluent at a flow rate of 18.0 mL/min and a linear gradient elution of 0–35% acetonitrile in 5–30 min. Collected appropriate fractions were concentrated using an evaporator and then lyophilized to give the polyamide **TH59** (16.9 mg, 7.5 × 10^{-3} mmol, 29% yield in 10 steps). Analytical HPLC: $t_R = 15.2$ min. ESI-TOF-MS m/z calcd for $C_{102}H_{129}N_{41}O_{20}{}^{2+}$ $[M + 2H]^{2+}$ 1124.0164 found 1124.0084 and calcd for $C_{102}H_{130}N_{41}O_{20}{}^{3+}[M + 3H]^{3+}$ 749.6800 found 749.6784.

Synthesis of Tandem Trimer Py–Im Polyamide TT59. Solid-phase synthesis was performed with 47.0 mg Fmoc-β-Wang resin (0.60 mmol/g). The resulting crude polyamide (33.6 mg) was purified by PU-2089 series HPLC (JASCO) using COSMOSIL 10 × 150 mm $5C_{18}$-MS-II Packed Column (Nacalai Tesque, Inc.) in 0.1% TFA in water with acetonitrile as the eluent at a flow rate of 3.0 mL/min and a linear gradient elution of 25 to 32.5% acetonitrile in 20 min. The collected fractions were treated as described above to give the polyamide **TT59** (8.0 mg, 2.4 × 10^{-3} mmol, 8.5% yield in 15 steps). Analytical HPLC: $t_R = 16.0$ min. ESI-TOF-MS m/z calcd for $C_{151}H_{187}N_{60}O_{31}{}^{3+}[M + 3H]^{3+}$ 1112.1634 found 1112.1313, calcd for $C_{151}H_{188}N_{60}O_{31}{}^{4+}[M + 4H]^{4+}$ 834.3739 found 834.3653 and calcd for $C_{151}H_{189}N_{60}O_{31}{}^{5+}[M + 5H]^{5+}$ 667.7006 found 667.7015.

Synthesis of the Florescent Probe TAMRA TH59. The polyamide **TH59** (1.3 mg, 5.8 × 10^{-4} mmol) and 5-carboxytetramethylrhodamine succinimidyl ester (0.60 mg, 1.1 × 10^{-3} mmol) were dissolved in DMF (300 μL) and DIEA (0.40 μL, 2.3 × 10^{-3} mmol), followed by mixing for 4 h at room temperature with shielding the light. The reaction mixture was purified by reversed-phase HPLC performed on a UV2075 HPLC UV/VIS detector and a PU-2080 plus series system (JASCO) using CHEMCOBOND 4.6 × 150 mm 5-ODS-H Column (Chemco) in 0.1% TFA in water with acetonitrile as the eluent at a flow rate of 1.0 mL/min and a linear gradient elution of 30–50% acetonitrile in 15 min. Appropriate fractions were collected under freeze-dry conditions to give the probe **TAMRA TH59** (1.6 mg) as a purple powder. Analytical HPLC: $t_R = 16.4$ min. ESI-TOF-MS m/z calcd for $C_{127}H_{149}N_{43}O_{24}{}^{2+}[M + 2H]^{2+}$ 1330.0875 found 1330.0630, calcd for $C_{127}H_{150}N_{43}O_{24}{}^{3+}[M + 3H]^{3+}$ 887.0607 found 887.0511 and calcd for $C_{127}H_{151}N_{43}O_{24}{}^{4+}[M + 4H]^{4+}$ 665.5474 found 665.5453.

Synthesis of the Florescent Probe TAMRA TT59. The polyamide **TT59** (0.93 mg, 2.8 × 10^{-4} mmol) and 5-carboxytetramethylrhodamine succinimidyl ester (0.35 mg, 6.6 × 10^{-4} mmol) were dissolved in DMF (200 μL) and DIEA (0.20 μL, 1.1 × 10^{-3} mmol), followed by mixing for 4 h at room temperature with shielding the light. The reaction mixture was purified as described above to give the probe **TAMRA TT59** (1.7 mg) as a purple powder. Analytical HPLC: $t_R = 17.0$ min. ESI-

TOF-MS m/z calcd for $C_{176}H_{207}N_{62}O_{35}{}^{3+}[M + 3H]^{3+}$ 1249.5436 found 1249.5127 and calcd for $C_{176}H_{208}N_{62}O_{35}{}^{4+}[M + 4H]^{4+}$ 937.4095 found 937.3922.

3.4.3 Absorption and Fluorescence Spectra

All spectra of fluorescent probes **TAMRA TH59** and **TAMRA TT59** were obtained in 7 mM sodium phosphate buffer (pH 7.0) containing 5.0% v/v DMF at room temperature. The UV–visible absorption spectra of 6.0 μM probes were obtained and steady state fluorescence spectra of 3.0 μM probes were measured. Steady state fluorescence spectra of 3.0 μM probes were obtained with an excitation wavelength of 540 nm in the absence or presence of 1.0 eq. or 2.0 eq. of ODN-1/2.

3.4.4 SPR Assays

The SPR assays were performed as described in previous reports [17b, 19]. Biotinylated DNA was immobilized to streptavidin-coated sensor chip SA at a flow rate of 10 μL/min to obtain the desired immobilization level (up to approximately 1400 RU rise). The assays were performed with HBS-EP (10 mM HEPES pH 7.4, 150 mM NaCl, 3 mM EDTA, 0.005% v/v Surfactant P20) with 0.1% DMSO at 25 °C. A series of sample solutions with various concentrations were prepared in HBS-EP buffer with 0.1% DMSO and injected at a flow rate of 20 μL/min. To calculate association rates (k_a), dissociation rates (k_d), and dissociation constants (K_D), data processing was performed with 1:1 binding with mass transfer model using BIA evaluation 4.1 program. The closeness of the fitting lines is described as χ^2 shown in the following equation. r_f is the fitted value at a given point, r_x is the experimental value at the same point, n is the number of data points, and p is the number of fitted parameters.

$$\chi^2 = \frac{\sum_1^n (r_f - r_x)^2}{n - p}$$

3.4.5 Telomeres Staining

Telomere Staining of HeLa 1.3 Cell Spreads with Fluorescent Polyamides [17b, 21]. HeLa 1.3 cells were maintained in DMEM medium containing 10% fetal bovine serum (FBS) at 37 °C (5% CO_2). HeLa 1.3 cells were blocked mitotically by adding 0.1 μg/mL nocodazole for 3 h. The cells were swollen by treatment with a hypotonic buffer (0.075 M KCl) for 15 min at room temperature and then fixed with a methanol/acetic acid (3:1) solution for 5 min. After centrifugation, cell pellet was again suspended in the methanol/acetic acid solution. The cell suspension was spread on coverslips and air-dried for 30 min. The cell spread coverslips were kept at 4 °C until use.

The air-dried spread coverslips were soaked in TEN buffer (10 mM Tris-HCl, pH 7.5, 1 mM EDTA and 100 mM NaCl) over night at 4 °C before use. For blocking, the spread coverslips were treated with 10% normal goat serum (NGS) in TEN buffer for 30 min at room temperature. After brief washing with TE buffer (10 mM Tris-HCl, pH 7.5, 1 mM EDTA), the spread coverslips were incubated with 10% NGS, 10 nM **TAMRA TH59** or 50 nM **TAMRA TT59** in DMF, and 0.5 μg/ml DAPI in TE buffer for 1 h at room temperature. After washing with TEN buffer (five times for 3 min), the spread coverslips were mounted in PPDI solution [10 mM HEPES pH 7.5, 100 mM KCl, 1 mM $MgCl_2$, 80% glycerol, 1 mg/mL paraphenylene diamine] [5] and the coverslips were sealed with a nailporish. Sectioning images were recorded with DeltaVision and projected ("Quick Projection" tool) without deconvolution to obtain telomeric signals and background signals.

Telomere Staining of Formaldehyde-fixed HeLa 1.3 Cells with Fluorescent Polyamides [17b]. For polyamide staining, the HeLa cells were grown on coverslips coated with polylysine. The cell coverslips were washed in phosphate-buffered saline (PBS) twice and fixed with 2% formaldehyde in PBS for 15 min at room temperature. The fixed cell coverslips were then treated with 50 mM glycine in PBS for 5 min, rinsed with PBS (twice for 5 min) and permeabilized with 0.5% Triton X-100 in PBS for 5 min. After briefly washing with HMK buffer (10 mM HEPES pH 7.5, 1 mM $MgCl_2$, 100 mM KCl) twice, the coverslips were soaked in HEN buffer (10 mM HEPES pH 7.5, 1 mM EDTA, 100 mM NaCl) and kept at 4 °C until use. For blocking, the cell coverslips were treated with 10% NGS in TE buffer (10 mM Tris-HCl pH 7.5, 1 mM EDTA) for 30 min at room temperature. After a brief rinse with TE buffer, the cells were incubated with 10% NGS, 10 nM TAMRA TH59 or 50 nM TAMRA TT59 in DMF, and 0.5 μg/mL DAPI in TE buffer at 37 °C for 1 h. After washing with TEN200 buffer (10 mM Tris-HCl pH 7.5, 1 mM EDTA, and 200 mM NaCl) (five times for 3 min). The mounting and subsequent image acquisitions were performed as described above. For quantification of the telomere signals in Fig. 3.4, they were extracted from the Fig. 3.3 images as described previously [16]. The telomere signals yielded by the polyamides were extracted based on threshold values using the Softworks software (Applied Precision). The maximum intensity values of signals in the extracted telomere regions were then used as telomere signals. For the background signals, 10 squares (10 pixels × 10 pixels) were randomly set outside the defined telomere signals and mean values of the signals in the squares were used as background signals.

Table S3.1 Sequence of the oligonucleotides. Underlined bases are binding sites of **TT59** and bold, italicized bases are mismatch sites for **TT59**

ODN-1	5′-GGTTAGGGTTAGGGTTAGGGTTAGG-3′
ODN-2	5′-CCTAACCCTAACCCTAACCCTAACC-3′
ODN-3	5′-biotin-GGTTAGGGTTAGGGTTAGGGTTAGGTTTTCCTAACCCTAACCCTAACCCTAACC-3′
ODN-4	5′-biotin-GGTTAGA***A***GTTAGGGTTAGGGTTAGGTTTTCCTAACCCTAACCCTAAC***T***CTAACC-3′
ODN-5	5′-biotin-GGTTAGGGTTAG***A***GTTAGGGTTAGGTTTTCCTAACCCTAAC***T***CTAACCCTAACC-3′

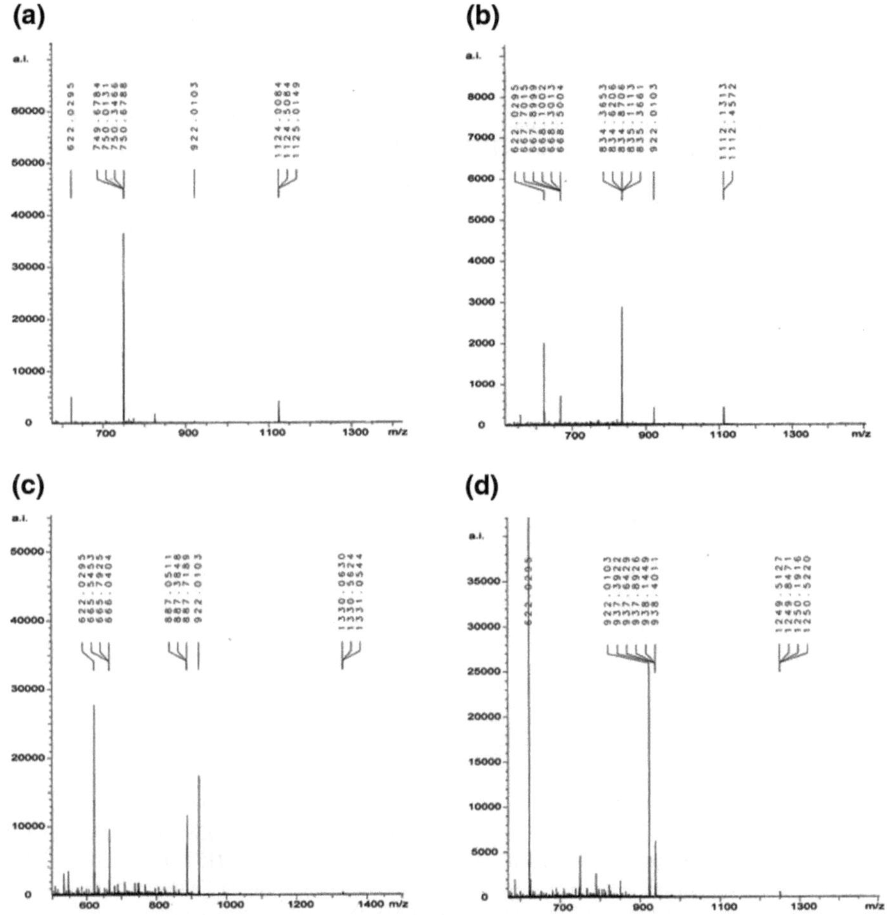

Fig. S3.1 ESI-TOF MS spectra of the polyamides **TH59**, **TT59**, **TAMRA TH59** and **TAMRA TT59** (**a**: **TH59**, **b**: **TT59**, **c**: **TAMRA TH59**, **d**: **TAMRA TT59**). Copyright 2015 The Royal Society of Chemistry

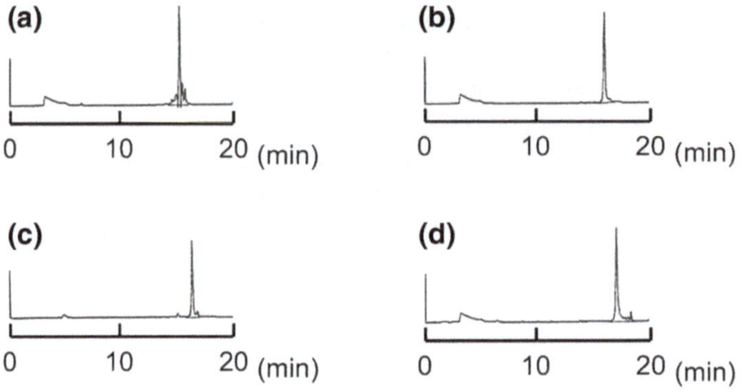

Fig. S3.2 Analytical HPLC profiles of the polyamides **TH59**, **TT59**, **TAMRA TH59** and **TAMRA TT59** (**a**: **TH59**, **b**: **TT59**, **c**: **TAMRA TH59**, **d**: **TAMRA TT59**). Copyright 2015 The Royal Society of Chemistry

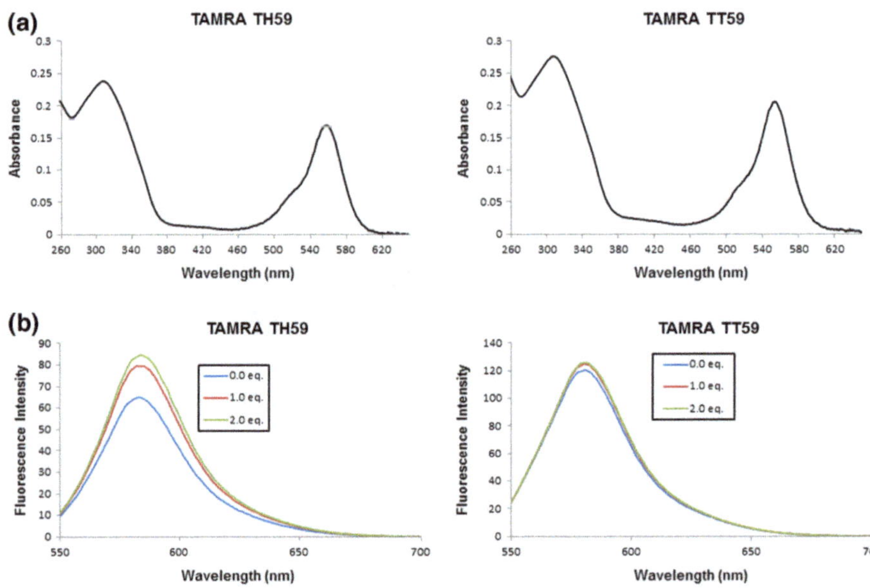

Fig. S3.3 **a** UV–visible absorption spectra of 6.0 μM probes **TAMRA TH59** and **TAMRA TT59** and **b** fluorescence spectra ($\lambda_{ex} = 540$ nm) of 3.0 μM probes **TAMRA TH59** and **TAMRA TT59** in the absence (blue) or presence of 1.0 equivalent (red) or 2.0 equivalent ODN-1/2 (green). Each spectrum was measured in 7 mM sodium phosphate buffer (pH 7.0) containing 5% v/v DMF. Copyright 2015 The Royal Society of Chemistry

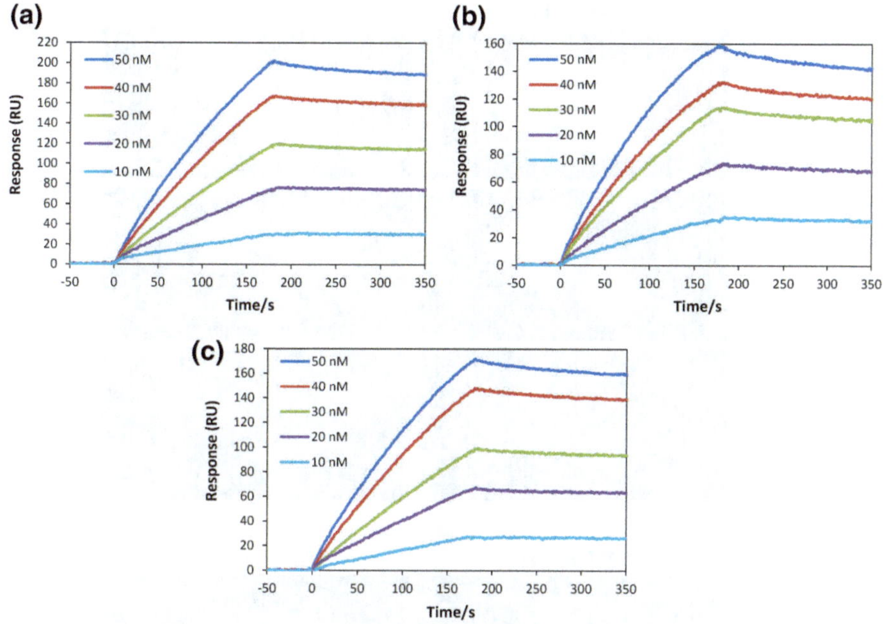

Fig. S3.4 SPR sensorgrams for the interaction of the polyamide **TH59** and **a** ODN-3, **b** ODN-4 or **c** ODN-5. Copyright 2015 The Royal Society of Chemistry

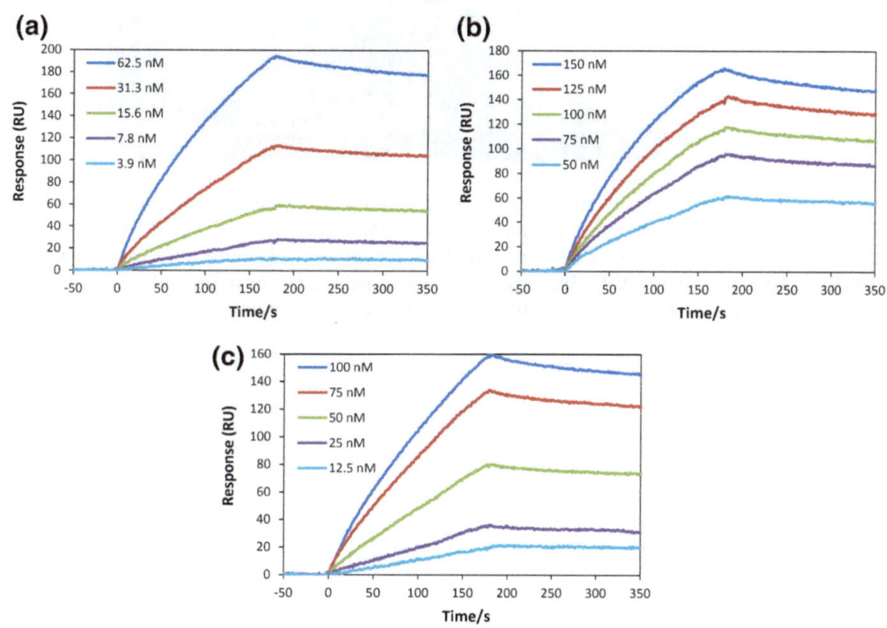

Fig. S3.5 SPR sensorgrams for the interaction of the polyamide **TT59** and **a** ODN-3, **b** ODN-4 or **c** ODN-5. Copyright 2015 The Royal Society of Chemistry

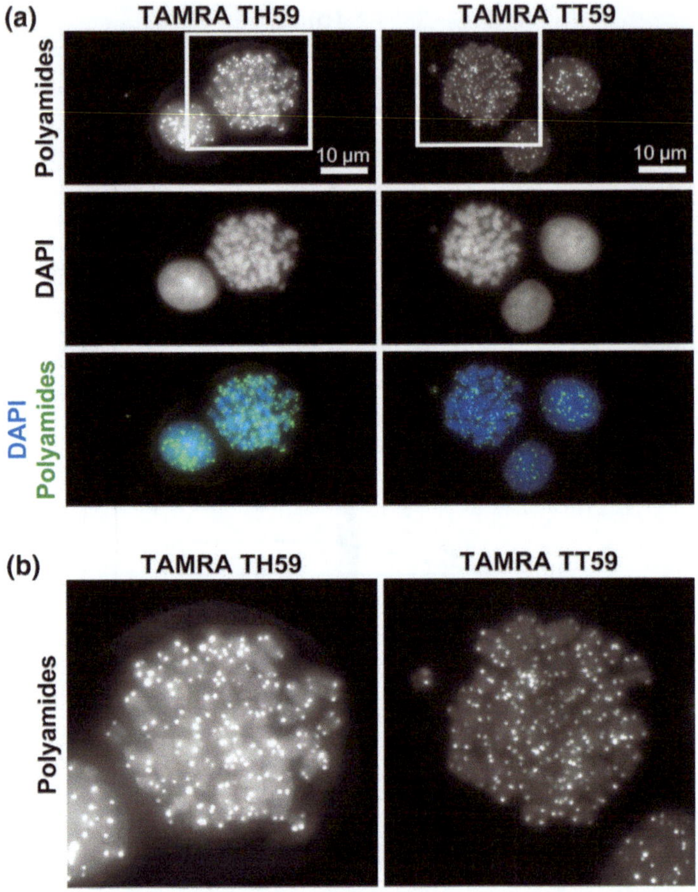

Fig. S3.6 Telomere staining of HeLa 1.3 cell spreads with fluorescent polyamides. **a** The HeLa 1.3 cell spread was stained with the fluorescent polyamide (first row) and DAPI (second row). The merged images are shown in the third row. The first and second columns show the results for **TAMRA TH59** and **TAMRA TT59**, respectively. Enlarged images of the boxed regions in panel (**a**) are shown in panel (**b**). Copyright 2015 The Royal Society of Chemistry

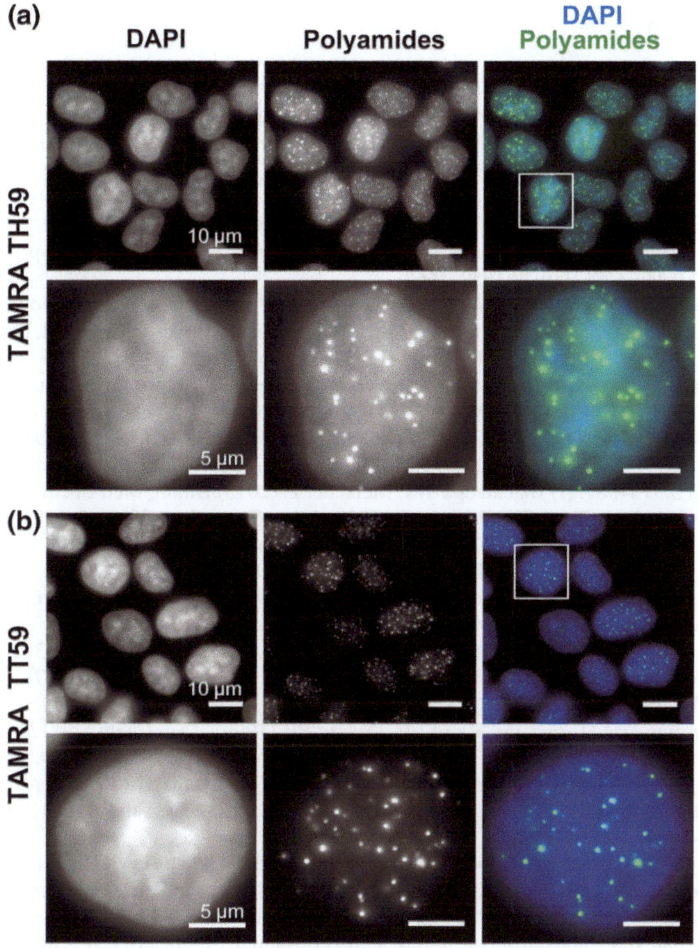

Fig. S3.7 Telomere staining of formaldehyde-fixed HeLa 1.3 cells with fluorescent polyamides. HeLa 1.3 cells were stained with DAPI (first column) and fluorescent polyamides (second column). The merged images are shown in the third column. Enlarged images of the boxed regions in the first row are shown in the second row. Cell images using **TAMRA TH59** and **TAMRA TT59** are shown in panels (**a**) and (**b**), respectively. Copyright 2015 The Royal Society of Chemistry

References

1. Blackburn EH (2010) Angew Chem Int Ed 49:7405–7421
2. (a) Nandakumar J, Cech TR (2013) Nat Rev Mol Cell Biol 14:69–82. (b) Biffi G, Tannahill D, McCafferty J, Balasubramanian S (2013) Nat Chem 5:182–186. (c) Doksani Y, Wu JY, de Lange T, Zhuang X (2013) Cell 155:345–356
3. Zakian VA (2012) Exp Cell Res 318:1456–1460
4. Smogorzewska A, de Lange T (2004) Annu Rev Biochem 73:177–208
5. Smogorzewska A, van Steensel B, Bianchi A, Oelmann S, Schaefer MR, Schnapp G, de Lange T (2000) Mol Cell Biol 20:1659–1668
6. (a) Boutorine AS, Novopashina DS, Krasheninina OA, Nozeret K, Venyaminova AG (2013) Molecules 18:15357–15397. (b) Lansdorp PM, Verwoerd NP, van de Rijke FM, Dragowska V, Little MT, Dirks RW, Raap AK, Tanke H (1996) J Hum Mol Genet 5:685–691
7. (a) Trauger JW, Baird EE, Dervan PB (1996) Nature 382:559–561. (b) White S, Szewczyk JW, Turner JM, Baird EE, Dervan PB (1998) Nature 391:468–471. (c) Dervan PB (2001) Bioorg Med Chem 9:2215–2235. (d) Dervan PB, Edelson BS (2003) Curr Opin Struc Biol 13:284–299. (e) Dervan PB, Doss RM, Marques MA (2005) Curr Med Chem: Anti-Cancer Agents 5:373–387. (f) Blackledge MS, Melander C (2013) Bioorg Med Chem 21:6101–6114
8. (a) Herman DM, Baird EE, Dervan PB (1998) J Am Chem Soc 120:1382–1391. (b) Swalley SE, Eldon E, Baird EE, Dervan PB (1999) J Am Chem Soc 121:1113–1120
9. Wurtz NR, Turner JM, Baird EE, Dervan PB (2001) Org Lett 3:1201–1203
10. (a) Trauger JW, Baird EE, Mrksich M, Dervan PB (1996) J Am Chem Soc 118:6160–6166. (b) Trauger JW, Baird EE, Dervan PB (1998) J Am Chem Soc 120:3534–3535
11. (a) Mrksich M, Parks ME, Dervan PB (1994) J Am Chem Soc 116:7983–7988. (b) de Clairac RPL, Geierstanger BH, Mrksich M, Dervan PB, Wemmer DE (1997) J Am Chem Soc 119:7909–7916
12. (a) Herman DM, Turner JM, Baird EE, Dervan PB (1999) J Am Chem Soc 121:1121–1129. (b) Chenoweth DM, Dervan PB (2010) J Am Chem Soc 132:14521–14529. (c) Morinaga H, Bando T, Takagaki T, Yamamoto H, Hashiya K, Sugiyama H (2011) J Am Chem Soc 133:18924–18930. (d) Li BC, Montgomery DC, Puckett JW, Dervan PB (2013) J Org Chem 78:124–133
13. (a) Bando T, Sugiyama H (2006) Acc Chem Res 39:935–944. (b) Xiao X, Yu P, Lim HS, Sikder D, Kodadek T (2007) Angew Chem Int Ed 46:2865–2868. (c) Jordan L, Meier JL, Yu AS, Korf I, Segal DJ, Dervan PB (2012) J Am Chem Soc 134:17814–17822. (d) Vaijayanthi T, Bando T, Pandian GN, Sugiyama H (2012) Chem BioChem 13:2170–2185. (e) Singh I, Wendeln C, Clark AW, Cooper JM, Ravoo BJ, Burley GA (2013) J Am Chem Soc 135:3449–3457. (f) Pandian GN, Taniguchi J, Junetha S, Sato S, Han L, Saha A, Anandhkumar C, Bando T, Nagase H, Vaijayanthi T, Taylor RD, Sugiyama H (2014) Sci Rep 4:3843
14. (a) Weyermann P, Dervan PB (2002) J Am Chem Soc 124:6872–6876. (b) Poulin-Kerstien AT, Dervan PB (2003) J Am Chem Soc 125:15811–15821. (c) Yamamoto M, Bando T, Morinaga H, Kawamoto Y, Hashiya K, Sugiyama H (2014) Chem Eur J 20:752–759
15. (a) Herman DM, Baird EE, Dervan PB (1999) Chem Eur J 5:975–983. (b) Kers I, Dervan PB (2002) Bioorg Med Chem 10:3339–3349. (c) Schaal TD, Mallet WG, McMinn DL, Nguyen NV, Sopko MM, John S, Parekh BS (2003) Nucleic Acids Res 31:1282–1291. (d) Sasaki S, Bando T, Minoshima M, Shinohara K, Sugiyama H (2008) Chem Eur J 14:864–870
16. Maeshima K, Janssen S, Laemmli UK (2001) EMBO J 20:3218–3228
17. (a) Kawamoto Y, Bando T, Kamada F, Li Y, Hashiya K, Maeshima K, Sugiyama H (2013) J Am Chem Soc 135:16468–16477. (b) Hirata A, Nokihara K, Kawamoto Y, Bando T, Sasaki A, Ide S, Maeshima K, Kasama T, Sugiyama H (2014) J Am Chem Soc 136:11546–11554. (c) Yamamoto M, Bando T, Kawamoto Y, Taylor R, Hashiya K, Sugiyama H (2014) Bioconjugate Chem 25:552–559. (d) Taylor RD, Kawamoto Y, Hashiya K, Bando T, Sugiyama H (2014) Chem Asian J 9:2527–2533
18. Rucker VC, Dunn AR, Sharma S, Dervan PB, Gray HB (2004) J Phys Chem B 108:7490–7494

19. Lacy ER, Le NM, Price CA, Lee M, Wilson WD (2002) J Am Chem Soc 124:2153–2163
20. Minoshima M, Bando T, Sasaki S, Fujimoto J, Sugiyama H (2008) Nucleic Acids Res 36:2889–2894
21. (a) Maeshima K, Laemmli UK (2003) Dev Cell 4:467–480. (b) Maeshima K, Yahata K, Sasaki Y, Nakatomi R, Tachibana T, Hashikawa T, Imamoto F, Imamoto N (2006) J Cell Sci 119:4442–4451

Chapter 4
Targeting 24 bp Within Telomere Repeat Sequences with Tandem Tetramer Pyrrole–Imidazole Polyamide Probes

Abstract Synthetic molecules that bind sequence-specifically to DNA have been developed for varied biological applications, including anticancer activity, regulation of gene expression, and visualization of specific genomic regions. Increasing the number of base pairs targeted by synthetic molecules strengthens their sequence specificity. Our group has been working on the development of pyrrole–imidazole polyamides that bind to the minor groove of DNA in a sequence-specific manner without causing denaturation. Recently, we reported a simple synthetic method of fluorescent tandem dimer polyamide probes composed of two hairpin moieties with a linking hinge, which bound to 12 bp in human telomeric repeats (5′–(TTAGGG)$_n$–3′) and could be used to specifically visualize telomeres in chemically fixed cells under mild conditions. We also performed structural optimization and extension of the target base pairs to allow more specific staining of telomeres. In the present study, we synthesized tandem tetramer polyamides composed of four hairpin moieties, targeting 24 bp in telomeric repeats, the longest reported binding site for synthetic, non-nucleic-acid-based, sequence-specific DNA-binding molecules. The novel tandem tetramers bound with a nanomolar dissociation constant to 24 bp sequences made up of four telomeric repeats. Fluorescently labeled tandem tetramer polyamide probes could visualize human telomeres in chemically fixed cells with lower background signals than polyamide probes reported previously, suggesting that they had higher specificity for telomeres. Furthermore, high-throughput sequencing of human genomic DNA pulled down by the biotin-labeled tandem tetramer polyamide probe confirmed its effective binding to telomeric repeats in the complex chromatinized genome.

Keywords Pyrrole–imidazole polyamides · Tandem tetramer polyamides · Solid-phase peptide synthesis · Telomere · Cellular imaging · High-throughput sequencing

---

This chapter is reprinted and modified with permission from "Targeting 24 bp within Telomere Repeat Sequences with Tandem Tetramer Pyrrole–Imidazole Polyamide Probes" Kawamoto, Y.; Sasaki, A.; Chandran, A.; Hashiya, K.; Ide, S; Bando, T.; Maeshima, K; Sugiyama, H. *J. Am. Chem. Soc.* **2016**, *138*, 14100–14107. Copyright 2016 American Chemical Society.

Y. Kawamoto, *Synthesis and Biological Evaluation of Pyrrole–Imidazole Polyamide Probes for Visualization of Telomeres*, Springer Theses,
https://doi.org/10.1007/978-981-13-6912-4_4

4.1 Introduction

Sequence-specific DNA binding molecules have potential as DNA-based therapeutics and diagnostics. Various kinds of programmable, sequence-specific DNA-binding molecules have been developed, including the clustered regularly interspaced short palindromic repeats (CRISPR)/CRISPR-associated caspase 9 (Cas9) systems [1], protein-based zinc fingers [2], transcription activator-like effector (TALE) [3], and nucleic-acid-based triplex-forming oligonucleotides (TFO) [4]. Chemically modified nucleic acid analogues such as peptide nucleic acids (PNA) can also function as TFOs [5]. Other synthetic, non-nucleic-acid-based sequence-specific DNA-binding molecules have been reported, including amidine-benzimidazole-phenyl (ABP)-based minor groove binders [6], DNA-binding fragments of transcription factors [7], threading polyintercalators [8] and minor groove-binding pyrrole–imidazole (Py–Im) polyamides [9]. Our group has been working on pyrrole–imidazole (Py–Im) polyamides first described by Dervan and coworkers. Py–Im polyamides are composed of *N*-methylpyrrole, *N*-methylimidazole, and aliphatic amino acids connected by amide bonds, and they bind sequence-specifically to the minor groove of double-stranded B-DNA by recognizing Watson–Crick base pairs. The binding rule is that antiparallel arrangement of Im/Py can bind to G·C base pairs, whereas Py/Py can bind to both A·T and T·A base pairs [9f, g]. Referred to as aliphatic amino acid residues, the C-terminal β-alanine tail and γ-aminobutyric acid or 2,4-diaminobutyric acid (Dab) turn moiety, called a γ-turn, also binds to A·T and T·A base pairs [10]. The γ-turn connects two aromatic peptides to form hairpin [11] and cyclic [12] structures, and the C-terminal β-alanine tail increases sequence specificity and binding affinity [13a]. In contrast, internal β-alanine can form β/β, β/Py and β/Im pairs as a substitution for Py rings and can relax the curvature of the polyamides to allow their efficient binding [13b]. Py–Im polyamides can be synthesized easily by machine-assisted solid-phase peptide synthesis (SPPS) [12d, 14] and have been used for many purposes; for example, visualization of specific sequences [15], pull-down and identification of binding regions [16], inhibition of gene expression [17], transcriptional activation [18, 19], and detection of specific sequences on a solid substrate [20]. To target a single site in the whole human genome, which comprises about 3 billion base pairs, a sequence of $\geq$16 bp must be recognized by DNA-binding molecules; therefore, many efforts have been made to extend the target base-pair sequence of such DNA ligands. Wilson and coworkers succeeded in selectively targeting a 10 bp sequence with ABP motifs [6a]. Mascareñas and coworkers have recently synthesized a DNA-binding peptide targeting 13 bp, which is composed of the DNA-binding fragments of two transcription factors connected by a hook [7]. Iverson and coworkers have developed threading polyintercalators, composed of naphthalene diimide intercalators and minor groove and major groove binding aliphatic linkers, which target 14 bp and 22 bp [8]. This 22 bp sequence was at that time the longest site recognized by a synthetic, non-nucleic-acid-based DNA binding molecule. Regarding Py–Im polyamides, a linear Py–Im polyamide monomer forming a homodimer with two molecules, and a cysteine-derived Py–Im polyamide dimer were both reported to bind to 16 bp sequences [21]. Furthermore, tandem hairpin motifs composed of several hairpin moieties with connecting hinge segments have also been demonstrated [22, 23].

Recently, our group has designed tandem hairpin Py–Im polyamides specific for the human telomere repetitive sequences 5′–(TTAGGG)$_n$–3′ [23]. The human telomere is composed of tandem repeats of 5′–TTAGGG–3′ and the protein complex called shelterin including TRF1, TRF2, Rap1, TIN2, TPP1 and POT1. The telomere protects the ends of the chromosomes from nucleolytic degradation and DNA recombination and thus is important in chromosome stability [24]. Human telomere DNA has a duplex region and a single-stranded 3′ overhang, which is suggested to form G-quadruplexes or t-loops [25]. In normal mammalian cells, the number of telomere repeats decreases with cell division and is related to the aging process and cancer [24a, c, f]. Therefore, the telomere length is an important biomarker in studies of these processes, and many techniques to visualize telomeres and to measure telomere length have been reported [15b]. Fluorescence in situ hybridization (FISH) using fluorescent PNA probes has been widely used, but this method normally denatures the DNA under the harsh conditions required for efficient probe binding to telomeres and can lead to destruction of telomere structures [26]. In contrast, Py–Im polyamides can bind to target sequences under mild conditions and thus can visualize target regions without destroying their structures [23e]. Based on a previous report of visualization of telomeres and estimation of their length in chemically fixed cells with Texas Red-labeled Py–Im polyamide probes [23a], our group developed a simple method for synthesis of Py–Im polyamide probes labeled with 5–carboxytetramethylrhodamine (TAMRA) that bound to 12 bp of the human telomere sequences. The probes were named tandem dimer **TDi59-A** and **TAMRA TDi59-A** (Fig. 4.1) and were synthesized from a fluorenylmethyloxycarbonyl (Fmoc) building block comprising the sequence from (R)-Dab to the N-terminus of a hairpin moiety [23b–d]. Costaining with the fluorescent Py–Im polyamide probe and anti-TRF1 antibody under mild conditions confirmed that fluorescently labeled **TDi59-A** could specifically visualize human telomeres in chemically fixed cells and that telomere length is related to the amount of shelterin [23a, b]. To decrease the background signals from nonspecific binding in telomere staining, we also developed **TAMRA TDi59-B**, containing hinge B lengthened by one methylene unit [23c], and tandem trimer **TAMRA TTri59-A**, whose target sequence was extended to 18 bp, the longest specific sequence reported for Py–Im polyamides [23d]. Furthermore the Py–Im polyamide probes can also visualize telomeres in tissue sections [23e].

In this study, we have synthesized, using an automatic solid-phase peptide synthesizer, novel tandem tetramer Py–Im polyamides **TTet59-A** and **-B** (Fig. 4.1) targeting 24 bp in the human telomere 5′–(TTAGGG)$_n$–3′ repeats. These Py–Im polyamides are composed of four hairpin units and three hinges, either A or B. To our knowledge, 24 bp is the new record for the longest binding site of synthetic, non-nucleic-acid-based, sequence-specific DNA-binding molecules. Surface plasmon resonance (SPR)

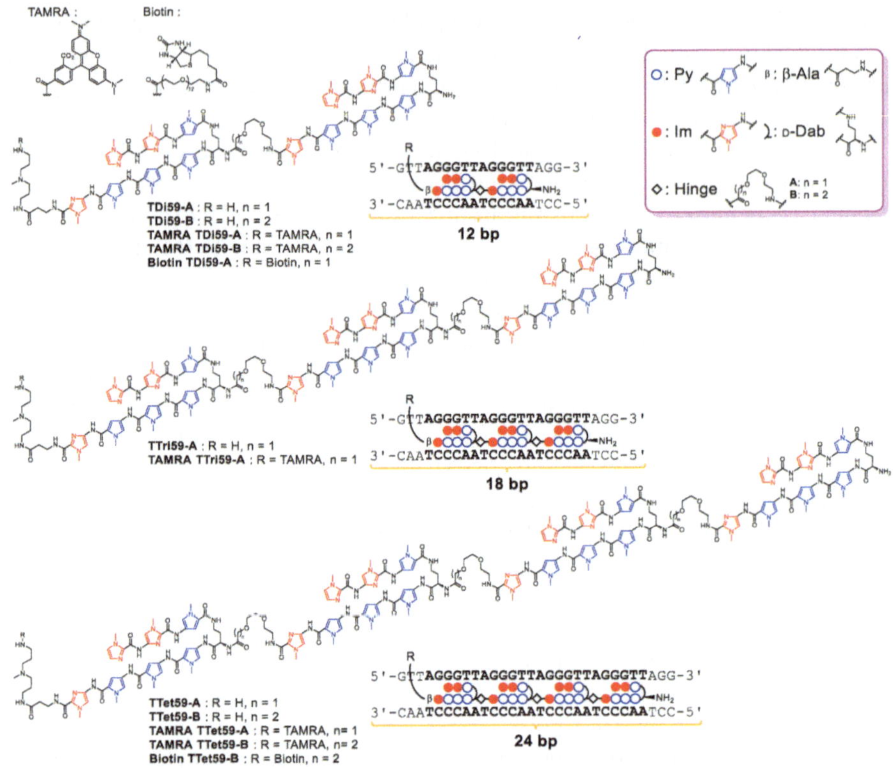

Fig. 4.1 Chemical structures and schematic representations of tandem dimer **TDi59–A** and **–B**, **TAMRA TDi59–A** and **–B** and **Biotin TDi59–A**, tandem trimer **TTri59–A** and **TAMRA TTri59–A**, and tandem tetramer **TTet59–A** and **–B** and **TAMRA TTet59–A** and **–B** and **Biotin TTet59–B** targeting for 12 bp, 18 bp, and 24 bp in human telomere sequence, respectively. Copyright 2016 American Chemical Society

analysis was performed to assess the binding affinity to four telomere repeats of the two tandem tetramer Py–Im polyamides and tandem trimer **TTri59-A**. After conjugation of **TTet59-A** and **-B** with TAMRA to obtain **TAMRA TTet59-A** and **-B**, we stained telomeres in chemically fixed cells to compare the new Py–Im polyamide probes with those previously reported and found that the background signals derived from nonspecific binding were much lower for the tetramers than for previously reported Py–Im polyamide probes. Finally, **TDi59-A** and **TTet59-B** were conjugated to biotin, allowing us to assess their binding to telomeric repeats in the whole genome using high-throughput sequencing technology [16f, g, 27].

4.2 Results and Discussion

4.2.1 Synthesis of Py–Im Polyamides

As shown in Scheme 4.1, tandem tetramer Py–Im polyamides **TTet59-A** and **-B** were synthesized based on the reported methodology, using building block **1** corresponding to Dab and three-ring Py and Im beside the N-terminal [23b–d]. In detail, machine-assisted Fmoc solid-phase peptide synthesis (SPPS) from Fmoc-β-Ala-Wang resin was performed. Fmoc-protected resin was treated with 20% piperidine/*N*,*N*-dimethylformamide (DMF) for deprotection, followed by coupling with the next Fmoc block (Fmoc-Py-OH, Fmoc-PyIm-OH [28], **1** [23b–d], Fmoc-mini-PEG or *N*-Fmoc-amido-dPEG2 Acid) activated with *N*,*N*-diisopropylethylamine (DIEA) and *N*,*N*,*N*,*N*-tetramethyl-O-(6-chloro-1H-benzotriazol-1-yl) uronium hexafluorophosphate (HCTU) in *N*-methyl-2-pyrrolidone (NMP). After these two procedures were repeated 19 times, the N-terminal Fmoc-protected amino group was deprotected with 20% piperidine and then cleaved from the resin with 3,3′-diamino-*N*-methyldipropylamine at 45 °C to produce **TTet59-A** and **-B** with 2.7 and 7.7% yield, respectively. The resulting Py–Im polyamides were coupled with 5-carboxytetramethylrhodamine or biotin succinimidyl ester in DIEA and DMF to produce the fluorescent probes **TAMRA TTet59-A** and **-B** and the biotinylated Py–Im polyamides **Biotin TDi59-A** and **Biotin TTet59-B**. Analytical high-performance liquid chromatography (HPLC) profiles and electrospray ionization time-of-flight mass (ESI-TOF-MS) spectrometry spectra are shown in Figs. S4.1 and S4.2, respectively.

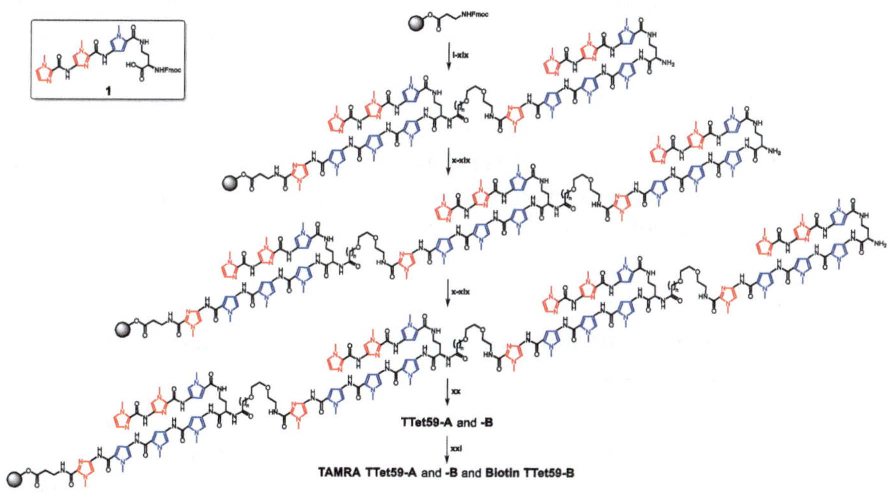

Scheme 4.1 Solid–phase synthesis of tandem tetramer Py–Im polyamides and the structure of building block **1**

Reagents and conditions: (i) 20% piperidine, DMF; (ii) Fmoc-PyIm-CO_2H, HCTU, DIEA, NMP; (iii) 20% piperidine, DMF; (iv) Fmoc-Py-CO_2H, HCTU, DIEA, NMP; (v) 20% piperidine, DMF; (vi) Fmoc-Py-CO_2H, HCTU, DIEA, NMP; (vii) 20% piperidine, DMF; (viii) **1**, HCTU, DIEA, NMP; (ix) 20% piperidine, DMF; (x) Fmoc-mini-PEG or *N*-Fmoc-Amido-dPEG$_2$ Acid, HCTU, DIEA, NMP; (xi) 20% piperidine, DMF; (xii) Fmoc-PyIm-CO_2H, HCTU, DIEA, NMP; (xiii) 20% piperidine, DMF; (xiv) Fmoc-Py-CO_2H, HCTU, DIEA, NMP; (xv) 20% piperidine, DMF; (xvi) Fmoc-Py-CO_2H, HCTU, DIEA, NMP; (xvii) 20% piperidine, DMF; (xviii) **1**, HCTU, DIEA, NMP; (xix) 20% piperidine, DMF; (xx) 3,3′-diamino-*N*-methyldipropylamine, 45 °C; (xxi) 5–carboxytetramethylrhodamine succinimidyl ester or EZ–Link NHS–PEG$_{12}$–Biotin, DIEA, DMF. Copyright 2016 American Chemical Society.

4.2.2 Comparative Analysis of Binding to Match Sequence

Binding affinities of tetramers **TTet59-A** and **-B** and tandem trimer **TTri59-A** to telomere repeats were assessed with SPR [23c, d, 29]. 5′-Biotinylated ODN-1 (5′-biotin-GGTTAGGGTTAGGGTTAGGGTTAGGGTTAGGTTTTCCTAACCCTAACCCTAACCCTAACCCTAACC-3′) containing four telomeric repeats was immobilized to a sensor chip through a biotin–streptavidin interaction, and then the trimer and tetramers were passed over the DNA on the sensor chip, and sensorgrams (shown in Fig. S4.3) and values (summarized in Table 4.1) were obtained. The underlined bases are the binding site of these three types of Py–Im polyamides. Tetramer **TTet59-A** showed strong binding affinity to ODN-1 (K_D = 7.5 nM), comparable to other types of Py–Im polyamides [9]. Trimer **TTri59-A** showed a slightly higher association constant and stronger binding affinity than **TTet59-A** (K_D = 6.1 nM). As suggested previously [23d], a major reason for this result is that the tetramer's higher steric hindrance influenced its accessibility to the match site. Of the three Py–Im polyamides, **TTet59-B** showed the strongest binding affinity (K_D = 4.0 nM), attributed to a much slower dissociation, suggesting that the hinge region prevented the polyamide's dissociation.

4.2.3 Visualization of Human Telomeres to Compare Py–Im Polyamide Probes

To compare the abilities of Py–Im polyamide probes to stain telomeres specifically, we doubly stained human HeLa 1.3 cell spreads with 4′,6-diamidino-2-phenylindole (DAPI) and five Py–Im polyamide probes: **TAMRA TDi59-A** and **-B**, **TAMRA TTri59-A**, and **TAMRA TTet59-A** and **-B** [23]. Cell spreads are used for clinical karyotyping tests and are prepared from colcemid-treated mitotic cells fixed with

Table 4.1 Binding affinities of polyamides **TTet59–A**, **TTet59–B** and **TTri59–A** against match sequence (ODN–1) calculated with SPR

	K_a ($M^{-1}S^{-1}$)	K_d (s^{-1})	K_D (M)
 TTet59-A	5.7×10^{-4}	4.3×10^{-4}	7.5×10^{-4}
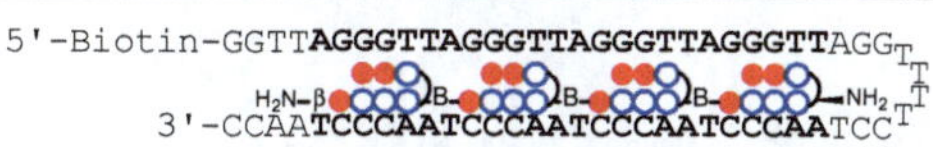 TTet59-B	4.1×10^{-4}	1.6×10^{-4}	4.0×10^{-4}
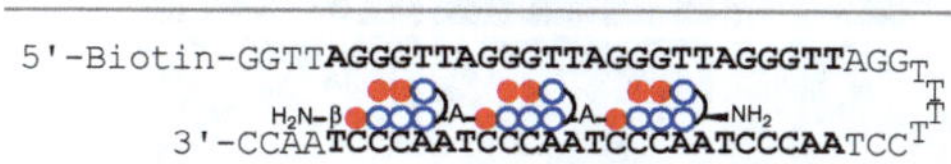 TTri59-A	7.9×10^{-4}	4.8×10^{-4}	6.1×10^{-4}

MeOH/AcOH. The resulting images of chromosomes stained with 75 nM Py–Im polyamide probes are shown in Figs. 4.2, S4.4a, and S4.5a. Chromosomal regions and nuclei were visualized with DAPI. Each Py–Im polyamide probe showed two foci at the ends of every chromosome, suggesting that all probes, including the new tandem tetramer probes **TAMRA TTet59-A** and **-B**, could stain telomeres specifically. As the number of targeted base pairs increased, the fluorescence of the polyamide probes became lower, presumably because increasing the size of the polyamide probes decreased the number of molecules binding to telomeres, but the background signals derived from nonspecific binding of polyamides decreased. When treating cells with **TAMRA TDi59-A** or **-B** or **TAMRA TTri59-A**, we could easily observe the shape of chromosomes from the background signals of polyamides, and stronger background signals for **TAMRA TDi59-A** in particular were detected from the whole chromosomes. However, for treatment with **TAMRA TTet59-A** or **-B**, the intensity of background signals was less, showing that extending the target base pair sequence significantly improves telomere specificity in cells. In addition, **TAMRA TTet59-B** had less background signal along the chromosomes than **TAMRA TTet59-A**, suggesting that the change of hinge from A to B is also effective in improving specificity; this was also shown by the comparison between **TAMRA TDi59-A** and **–B** [23c]. For quantitative evaluation of the telomere staining, we compared relative signal-to-background (S/B) ratios between polyamide probes (Fig. 4.2b). Each polyamide signal at chromosome ends was extracted, and divided by background signal at its neighboring chromosomal region. **TAMRA TTet59-B** had the highest relative S/B ratio. These results demonstrated that increasing the number of target base pairs to 24 and changing the hinge from A to B suppressed the nonspecific binding and allowed us to observe telomeres in HeLa 1.3 cell spreads with high specificity.

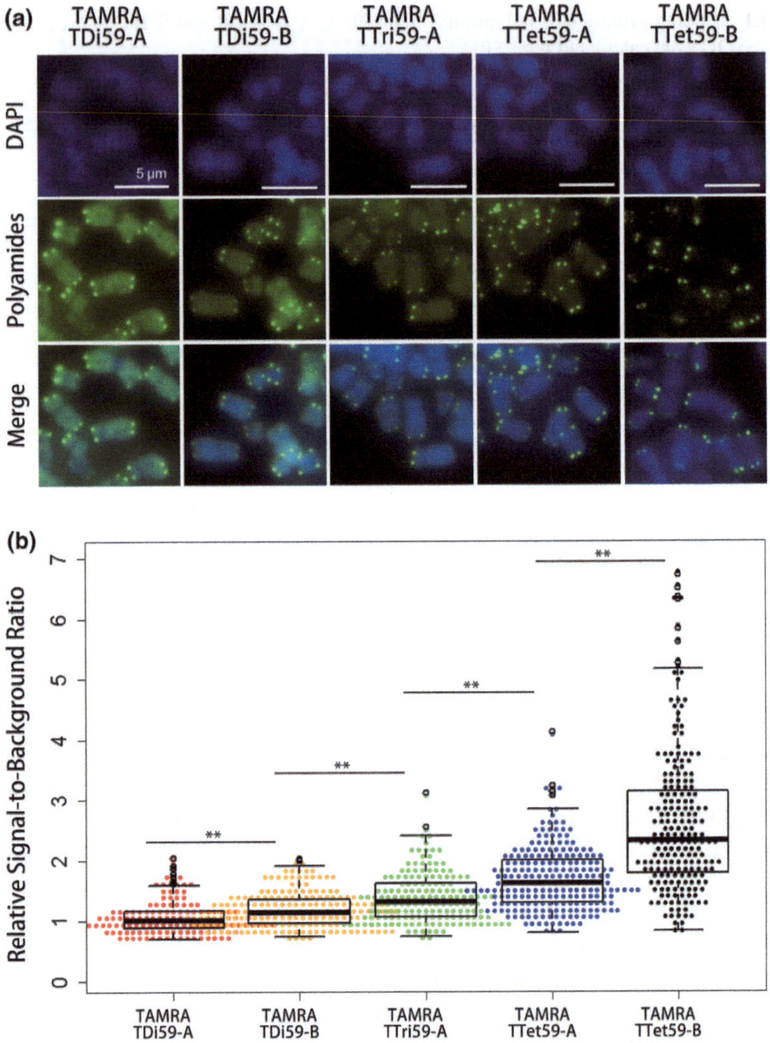

Fig. 4.2 Telomere staining of HeLa 1.3 cell spreads with 75 nM fluorescent Py–Im polyamide probes. **a** The images of HeLa 1.3 cell spreads stained with DAPI (first row) and the fluorescent polyamide probes (second row). The merged images are shown in the third row. Original images are shown in Fig. S4.4a. **b** Signal-to-background (S/B) ratios of telomeric foci in images. Dot plots and box plots are based on the images shown in Fig. S4.4a (for each group, number of dots = 167–231). All the values were expressed relative to the median value of **TAMRA TDi59-A**. $^{**}P < 0.01$, Wilcoxon rank sum test. Copyright 2016 American Chemical Society

Images of telomere staining of HeLa 1.3 cells fixed in formaldehyde (Figs. 4.3, S4.4b and S4.5b), a commonly used fixation method in cell biology, and the quantitative analysis of S/B ratios (Fig. 4.3b) suggested that the results were consistent with those of cell spreads staining. Many sharp foci from the telomeric regions were observed in the nuclei of cells [23a, b]. As reported previously, **TAMRA TDi59-B** and **TAMRA TTri59-A** showed lower background signals than **TAMRA TDi59-A** [23c, d]. The background signals of the two tandem tetramer probes **TAMRA TTet59-A** and **-B** were lower and their S/B ratios were higher than the other three probes. As can be seen from the polyamide signals of **TAMRA TTet59-A** and **-B** shown in Fig. 4.3, the shapes of the whole nuclei were less distinct. In the images of formaldehyde-fixed cells, **TAMRA TTet59-B** also showed the highest S/B ratio. These results also demonstrated that **TAMRA TTet59-A** and **-B** could visualize telomeres with higher specificity in the fixed cells or formaldehyde-fixed cells and that **TAMRA TTet59-B** targeting 24 bp and composed of four hairpins and three B hinges had the highest telomere specificity of the fluorescent Py–Im polyamide-based telomere probes.

4.2.4 High-Throughput Sequencing of the Binding Sites of Py–Im Polyamide Probes for Telomeric Repeats

Images of chromosome spreads treated with the Py–Im polyamide probes (Fig. 4.2) suggested that they specifically stained telomeres. To assess their efficient binding to telomeric repeats in a biologically active, histone-packed, chromatinized genome, we performed high-throughput sequencing of DNA bound to **Biotin TDi59-A** and **Biotin TTet59-B**, based on the reported procedure [16f, g]. The workflow is shown in Scheme 4.2. In brief, nuclei were extracted from BJ fibroblast cells and then incubated with 400 nM **Biotin TDi59-A** or **Biotin TTet59-B** for 16 h at 4 °C. After this, the extracted nuclei were treated with micrococcal nuclease to produce mononucleosomes to generate the optimally sized DNA for construction of a sequencing library. Nuclear proteins were digested with proteinase K, and then DNA bound to the biotinylated Py–Im polyamide probes was extracted through a biotin–streptavidin interaction on streptavidin–coated magnetic beads (C1 beads). The extracted and purified DNA was ligated with sequencing platform-specific adapters and amplified with a polymerase chain reaction to obtain sufficient amounts for a sequencing library. The qualified libraries were sequenced using Ion Proton Sequencing system. Sequenced and filtered reads were aligned to human reference genome hg38. MACS peaks were called to identify the differentially enriched regions compared with control data [30a].

Figure S4.6 shows a representative result indicating the enriched regions in chromosomes 1 and 18, and a close-up of the enriched telomeric repeats from a non-treated control sample and after **Biotin TDi59-A** and **Biotin TTet59-B** treatment is shown in Fig. 4.4. In both left and right arm termini of chromosomes 1 and 18, we observed

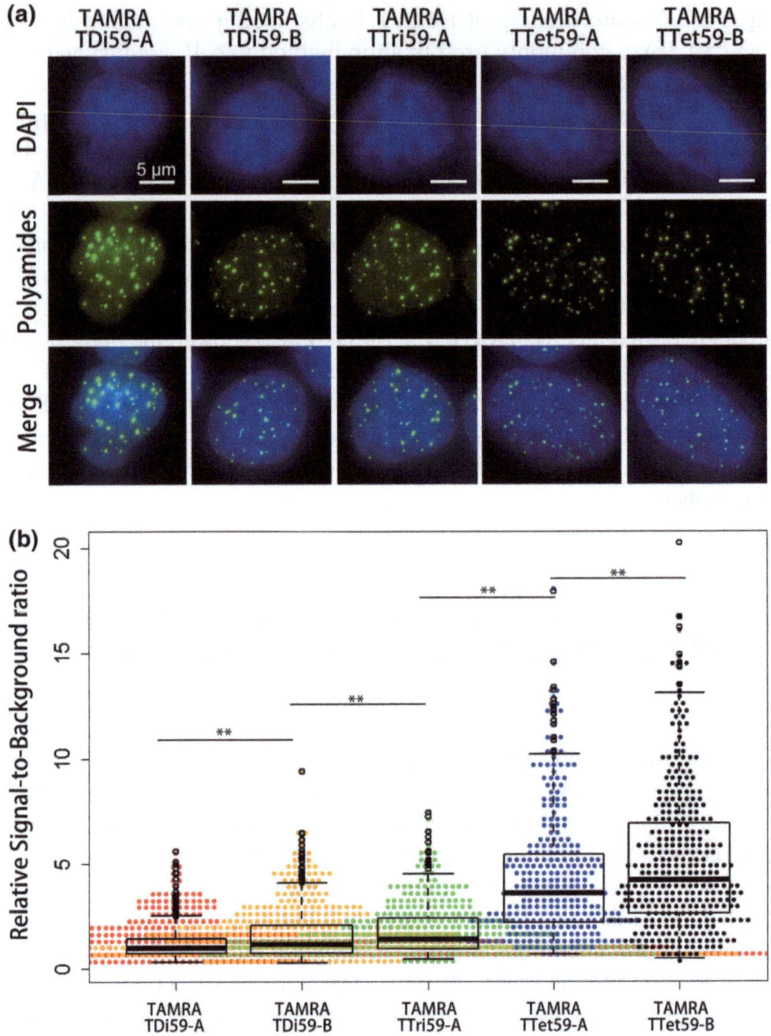

Fig. 4.3 Telomere staining of HeLa 1.3 cells with 15 nM fluorescent Py–Im polyamide probes. **a** The images of HeLa 1.3 cells stained with DAPI (first row) and fluorescent polyamide probes (second row). The merged images are shown in the third row. Original images are shown in Fig. S4.4b. **b** Signal-to-background (S/B) ratios of telomeric foci in images. Dot plots and box plots are based on the images shown in Fig. S4.4b (for each group, number of dots = 295–606). All the values were expressed relative to the median value of **TAMRA TDi59-A**. $^{**}P < 0.01$, Wilcoxon rank sum test. Copyright 2016 American Chemical Society

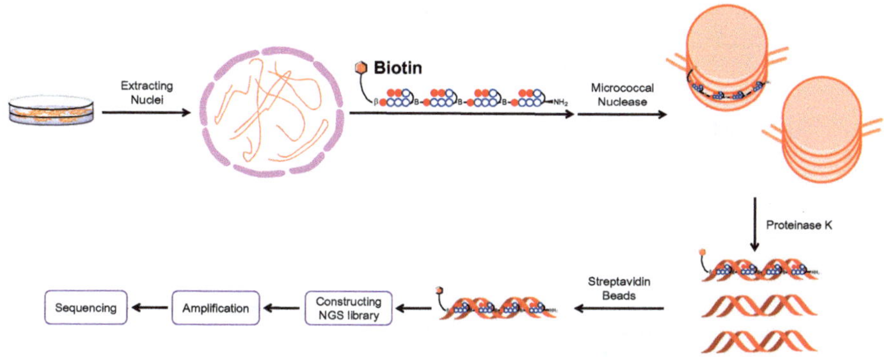

Scheme 4.2 Extraction and high–throughput sequencing of genomic DNA Bound by biotinylated Py–Im polyamide probes Copyright 2016 American Chemical Society

significant enrichments with both polyamides compared with the control. The enrichment was consistent with telomeric repeat sequences (TTAGGG/CCCTAA) of the reference genome, suggesting that we could successfully validate the binding of **Biotin TDi59-A** and **Biotin TTet59-B** to telomeric repeats with high-throughput sequencing. Other mild enrichments appeared to be the results of partial or non–specific binding.

The enrichments in the termini of each chromosome are shown in Figure S4.6. Duplicate experiments ensured the consistency of enrichments in chromosomal termini. However, we could not quantify the enrichment level on Chr 3R, 5R, 6L, 6R, 8L, 8R, 9R, 11L, 13L, 14L, 14R, 15L, 16R, 17L, 19L, 20L, 20R, 21L and 22L (L-left arm, R-right arm) chromosomal termini. This may be because of lacking telomeric repeat sequences information on the above mentioned chromosomes in the human reference genome hg38 data. Even though chromosomes 13, 14, 15, 21 and 22 lack the telomeric repeats information at the left termini presumably because of rDNA whose sequences are not included in the human reference genome hg38, we could observe enrichments from ~16,000,000, ~16,000,000, ~17,000,000, ~5,010,000 and ~10,510,000 bp, respectively. These results show that **Biotin TDi59-A** and **Biotin TTet59-B** could effectively recognize and bind to telomeric repeats.

To identify high affinity binding motif of **Biotin TDi59-A** and **Biotin TTet59-B** from the enrichments, we used the Homer motif analysis program with the enriched peaks [30b]. One of the highly enriched motifs corresponded to telomeric 5′–(TTAGGG)$_n$–3′ repeats with a p value of e^{-8595} and e^{-100} for **Biotin TDi59-A** and **Biotin TTet59-B**, respectively (Fig. 4.4c). This result also suggests efficient binding of **Biotin TDi59-A** and **Biotin TTet59-B** to telomeres.

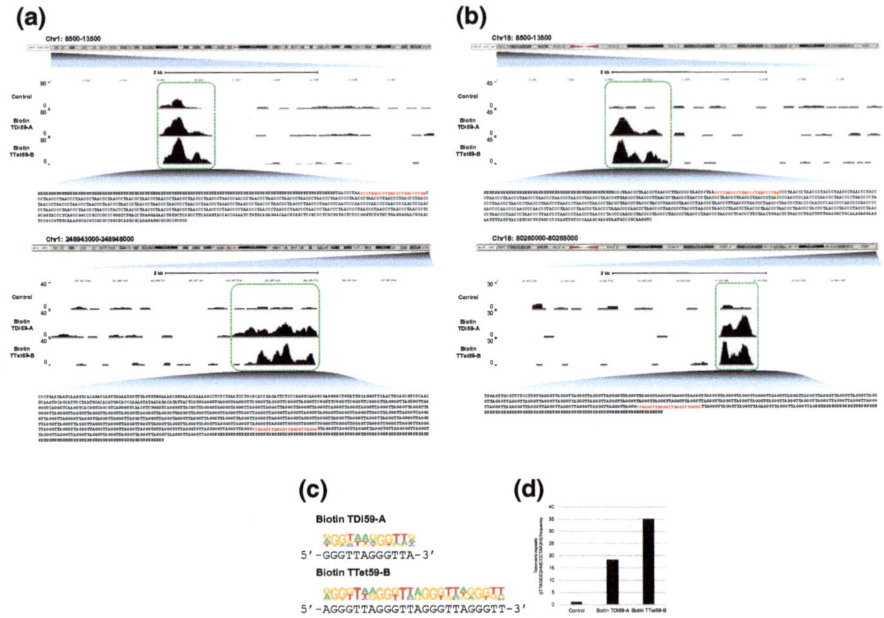

Fig. 4.4 Assessment of efficient binding of **Biotin TDi59–A** and **Biotin TTet59–B** to telomeric repeats with high–throughput sequencing. Mapping of the biotinylated Py–Im polyamide probes binding and enriched sites in both termini of chromosomes 1 and 18 are shown in (**a**) and (**b**) respectively. **c** Binding motifs of the probes corresponding to telomeric repeats identified in human genomic enriched sequences. **d** The ratio of sequenced reads containing ≥4 telomeric repeats to whole reads as normalized to that from the control. Copyright 2016 American Chemical Society

Finally, to compare the recognition of telomeric repeats by **Biotin TDi59-A** with that by **Biotin TTet59-B**, the ratio of reads containing ≥ 4 TTAGGG/CCCTAA repeats to whole sequenced reads was calculated. As shown in Fig. 4.4d, the ratio for **Biotin TDi59-A** was 18.2 times higher than the nontreated control, showing that **Biotin TDi59-A** could enrich telomeric DNA. The ratio for **Biotin TTet59-B** was 1.9 times higher than that for **Biotin TDi59-A**, suggesting that **TTet59-B** exhibited better recognition ability toward telomeric repeats. This result is also consistent with the higher telomere-specific staining of **TAMRA TTet59-B**.

High-throughput sequencing confirmed the binding of **Biotin TDi59-A** and **Biotin TTet59-B** to telomeric repeats in the whole genome. In contrast, foci of fluorescently labeled Py–Im polyamide probes were specifically detected at the end of the chromosomes (Fig. 4.2). The principal reason for this is that accumulation of the Py–Im polyamides on long telomeric repeats in telomere regions was higher than that in nonspecific regions, and therefore much stronger fluorescence was detected at the termini of chromosomes than from other chromosomal regions.

4.3 Conclusion

In this study, we have synthesized novel tandem tetramer Py–Im polyamides composed of 4 hairpins and 3 hinges that targeted 24 bp of the human telomere sequences. This is a new record for the longest binding site of synthetic, non-nucleic-acid-based, sequence-specific DNA-binding molecules. SPR analysis revealed that tandem tetramers could bind to four telomeric repeats with nanomolar K_D values. TAMRA-labeled tandem tetramer Py–Im polyamide probes could stain clearly the telomere foci in chemically fixed cells. Compared with previously reported tandem dimer and trimer Py–Im polyamides, the tandem tetramers had higher specificity for telomeres in cells, and **TAMRA TTet59-B** containing hinge B had the highest specificity. Furthermore, high-throughput sequencing of chromatin pulled down by biotin-labeled Py–Im polyamide probes confirmed their high recognition and binding toward telomeric DNA. To the best of our knowledge, this is the first report of high-throughput sequencing of DNA pulled down by a synthetic DNA binder targeting human telomeres. To allow the application of these tetramers, our group continues to optimize further their chemical structures and to develop new functions.

4.4 Supporting Information

4.4.1 Materials

HPLC analysis for chasing each solution–phase reaction and checking each product was performed on a PU–2089 plus series system (JASCO) using CHEMCOBOND 4.6 mm × 150 mm 5–ODS–H Column (Chemco Plus Scientific Co., Ltd.) in 0.1% TFA in water with acetonitrile as the eluent at a flow rate of 1.0 mL/min and a linear gradient elution of 0–50% acetonitrile in 20 min with detection at 254 nm. Purification of Py–Im polyamides were done with reversed–phase HPLC performed on a UV2075 HPLC UV/VIS detector and a PU–2080 plus series system (JASCO) using CHEMCOBOND 4.6 mm × 150 mm 5–ODS–H Column in 0.1% TFA in water with acetonitrile as the eluent at a flow rate of 1.0 mL/min. Collected fractions were lyophilized and then analyzed by ESI–TOF–MS (Bruker). For the collect mass measurement, external mass calibration (m/z: 622.0290 and 922.0098), purchased from Agilent Technologies, was used. Fmoc-mini-PEG and *N*-Fmoc-Amido-dPEG$_2$ Acid were purchased from Peptide International. Fmoc-Py-CO_2H, DMF used for solid–phase synthesis, NMP, trifluoroacetic acid (TFA), and piperidine were from Wako. HCTU was from Peptide Institute, Inc. Fmoc-β-Ala-Wang resin (0.60 mmol/g, 100–200 mesh) was from Novabiochem. DIEA was from Nacalai Tesque, Inc. DMF used for solution–phase synthesis and for telomere–staining and acetonitrile were from Kanto Chemical Co., Inc. 3,3′-diamino-*N*-methyldipropyl-amine was from Tokyo Chemical Industry Co., Ltd. 5–carboxytetramethylrhodamine succinimidyl ester and EZ–Link NHS–PEG$_{12}$–Biotin were from Thermo Fisher Scientific Inc.

The Fmoc units **1** [23b, c] and Fmoc-PyIm-CO_2H [28a] were synthesized using the previous method. **TAMRA TDi59–B** was purchased from HiPep Laboratories. **TAMRA TDi59–A, TTri59–A** and **TAMRA TTri59–A** were synthesized following the previous methodology [23c, d]. Solid–phase peptide synthesis was performed on a PSSM–8 (Shimadzu). Concentrations of Py–Im Polyamides were calculated with a Nanodrop ND–1000 spectrophotometer (Thermo Fisher Scientific Inc.) using an extinction coefficient of 9900 $M^{-1}cm^{-1}$ per one pyrrole or imidazole moiety at λ_{max} near 310 nm. The SPR assays were performed using a BIACORE X instrument (GE Healthcare) with 5′–biotinylated DNA oligonucleotides (JBioS), streptavidin–coated sensor chip SA (GE Healthcare) and HBS–EP buffer (GE Healthcare). Cell images were recorded with DeltaVision (Applied Precision). HeLa 1.3 cells were generous gifts of Dr. T. de Lange (Rockefeller University) and BJ fibroblast cells were purchased from ATCC. DMEM medium and streptavidin–coated magnetic beads were purchased from Invitrogen. Normal goat serum (NGS) was from Millipore. Para-phenylene diamine, Triton X–100 and protease inhibitor cocktail were from Sigma. DAPI was from Roche. MNase was from TaKaRa Bio Inc. Nonidet P–40 was from Nacalai Tesque, Inc. PCR purification kit was from Qiagen. Quality and quantity check of sequencing library was performed with Agilent 2100 Bioanalyzer and High sensitivity BioAnalyzer kit (Agilent Technologies). High throughput sequencing was performed with Ion Proton^{TM} Sequencer (Thermo Fisher Scientific Inc.). All other solvents and materials were from standard suppliers (highest quality available).

4.4.2 Synthetic Details of Py–Im Polyamides

General Procedures of Solid–Phase Synthesis of Tandem Tetramer Py–Im Polyamides. Following the reported condition [23d], each Fmoc solid–phase peptide synthesis (SPPS) of tandem tetramers was performed with a computer–assisted operation system under low humidity (<20%). Fmoc units in each step were as follows: Fmoc-Py-CO_2H (77 mg, 0.21 mmol), Fmoc-PyIm-CO_2H (70 mg, 0.14 mmol), Fmoc-D-Dab(ImImPy)-OH (**1**) (70 mg, 0.10 mmol), Fmoc-mini-PEG (81 mg, 0.21 mmol) and *N*-Fmoc-Amido-dPEG_2 Acid (84 mg, 0.21 mmol). Before SPPS, each Fmoc unit was dissolved in NMP (1.0 mL) with HCTU and the amount of HCTU for Fmoc-Py-CO_2H, Fmoc-PyIm-CO_2H, **1**, Fmoc-mini-PEG and *N*-Fmoc-Amido-dPEG_2 Acid was 88 mg (0.21 mmol), 60 mg (0.15 mmol), 40 mg (0.097 mmol), 88 mg (0.21 mmol) and 88 mg (0.21 mmol), respectively. Fmoc-β-Ala-Wang resin swelled with NMP was loaded in the reaction vessel.

Procedures were as follows: twice deblocking for 4 min with 20% piperidine/DMF (500 μL), addition of 10% DIEA/NMP (364 μL, DIEA was 0.21 mmol) to Fmoc unit-HCTU mixture for activation, coupling for 60 min. Five times washing with DMF were performed during each step. After the last coupling, the deblocking of amino group at γ-turn was performed. All couplings were carried out with a single-coupling cycle. All lines were purged with solution transfers and bubbled by nitrogen gas for stirring the resin.

After SPPS, Py–Im polyamides on resin were cleaved with 3,3'-diamino-*N*-methyldipropylamine at 45 °C for 3 h. The resin was removed by filtration and washed thoroughly with dichloromethane, and the filtrate was concentrated *in vacuo*. The residue was dissolved in 1.0–2.0 mL dichloromethane-methanol mixture and then more than 10-fold volume of diethyl ether was added, followed by the centrifuge and removal of the supernatant. This process was repeated a few times until white precipitation was obtained. The obtained crude products were purified by reversed–phase HPLC in 0.1% TFA in water with acetonitrile as the eluent at a flow rate of 1.0 mL/min and a linear gradient elution of 25–55% acetonitrile in 40 min. Collected fractions were lyophilized to obtain the objective compounds.

Synthesis of TTet59–A. Solid–phase synthesis was performed with 42.0 mg Fmoc–β–Wang resin (0.60 mmol/g) and finally **TTet59–A** (3.0 mg, 6.8×10^{-4} mmol, 2.7% yield) was obtained as a pale yellow solid. Analytical HPLC: t_R = 16.9 min. ESI–TOF–MS *m/z* calcd for $C_{200}H_{246}N_{79}O_{42}{}^{5+}$ $[M + 5H]^{5+}$ 885.1903 found 885.1820.

Synthesis of TTet59–B. Solid–phase synthesis was performed with 43.0 mg Fmoc–β–Wang resin (0.60 mmol/g) as a pale yellow solid. The resulting crude polyamide was purified as described to afford **TTet59–B** (8.9 mg, 2.0×10^{-3} mmol, 7.7% yield). Analytical HPLC: t_R = 17.1 min. ESI–TOF–MS *m/z* calcd for $C_{203}H_{252}N_{79}O_{42}{}^{5+}$ $[M + 5H]^{5+}$ 893.5997 found 893.5936.

Synthesis of TAMRA TTet59–A. TTet59–A (0.9 mg) and 5–carboxytetramethyl-rhodamine succinimidyl ester (0.2 mg) were dissolved in DMF (250 μL) and DIEA (0.40 μL), followed by mixing at room temperature with shielding the light. After checking this reaction had been finished, the reaction mixture was purified by reversed–phase HPLC in 0.1% TFA in water with acetonitrile as the eluent at a flow rate of 1.0 mL/min and a linear gradient elution of 30 to 70% acetonitrile in 30 min, followed by lyophilization of the collected fractions to afford **TAMRA TTet59–A** (1.1 mg) as a purple powder. Analytical HPLC: t_R = 18.2 min. ESI–TOF–MS *m/z* calcd for $C_{225}H_{267}N_{81}O_{46}{}^{6+}$ $[M + 6H]^{6+}$ 806.5169 found 806.5205.

Synthesis of TAMRA TTet59–B. TTet59–B (2.6 mg) and 5–carboxytetramethyl-rhodamine succinimidyl ester (0.9 mg) were dissolved in DMF (272 μL) and DIEA (0.60 μL). The reaction and purification were performed as described above to afford **TAMRA TTet59–B** (2.0 mg) as a purple powder. Analytical HPLC: t_R = 18.3 min. ESI–TOF–MS *m/z* calcd for $C_{228}H_{274}N_{81}O_{46}{}^{7+}$ $[M + 7H]^{7+}$ 697.4508 found 697.4565.

Synthesis of Biotin TDi59–A. TDi59–A (1.9 mg) and EZ–Link NHS–PEG_{12}–Biotin 0.9 mg were dissolved in DMF (159 μL) and DIEA (0.30 μL). The reaction and purification were performed as described above to afford **Biotin TDi59–A** (0.4 mg) as a white solid. Analytical HPLC: t_R = 17.4 min. ESI–TOF–MS *m/z* calcd for $C_{139}H_{198}N_{44}O_{35}S^{4+}$ $[M + 4H]^{4+}$ 768.8691 found 768.8644.

Synthesis of Biotin TTet59–B. TTet59–B (0.8 mg) and EZ–Link NHS–PEG_{12}–Biotin 0.1 mg were dissolved in DMF (101 μL) and DIEA (0.20 μL). The reaction and purification were performed as described above to afford **Biotin TTet59–B** (0.5 mg) as a white solid. Analytical HPLC: $t_R = 19.0$ min. ESI–TOF–MS *m/z* calcd for $C_{240}H_{321}N_{82}O_{57}S^{7+}$ $[M + 7H]^{7+}$ 756.4918 found 756.4987.

4.4.3 SPR Assays

The SPR assays were performed as described in previous reports [23c, d]. Biotinylated DNA was immobilized to streptavidin–coated sensor chip at a flow rate of 10 μL/min to obtain the desired immobilization level (up to approximately 1000 RU rise). The assays were performed with HBS–EP (10 mM HEPES pH 7.4, 150 mM NaCl, 3 mM EDTA, 0.005% v/v Surfactant P20) containing 0.1% DMSO at 25 °C. A series of sample solutions with various concentrations were prepared in HBS–EP buffer containing 0.1% DMSO and injected at a flow rate of 20 μL/min. When preparing the solution of **TTet59–A**, its portion was dissolved in DMSO and the concentration was determined to be 467 μM by checking its absorbance as mentioned above. DMSO was added to the solution to dilute it to 400 μM, followed by pouring 1 μL of the resulting solution into 1 mL HBS-EP buffer. The 400 nM solution was diluted to each concentration by addition of the buffer containing 0.1% DMSO. Solutions of other compounds were prepared in the same method. To calculate association rates (k_a), dissociation rates (k_d), and dissociation constants (K_D), data processing was performed with 1:1 binding with mass transfer model using BIAevaluation 4.1 program.

4.4.4 Telomeres Staining and Calculation of S/B Ratios

HeLa 1.3 Cell Spreads Treated with Fluorescent Py–Im Polyamide Probes [23, 31]. HeLa 1.3 cells were maintained in DMEM medium containing 10% fetal bovine serum (FBS) at 37 °C (5% CO_2). HeLa 1.3 cells were blocked mitotically by adding 0.2 μg/mL colcemid and incubated overnight. The cells were then swollen by treatment with hypotonic buffer (0.075 M KCl) for 15 min at room temperature, fixed with methanol/acetic acid (3:1) solution for 5 min. After centrifugation, cell pellet was again suspended in the methanol/acetic acid (3:1) solution. The cell suspension was spread on coverslips and air–dried at room temperature for 5 min and then at 60 °C for 30 min. The cell spread coverslips were kept at 4 °C until use.

The dried spread coverslips were soaked in HEN buffer (10 mM HEPES pH 7.5, 1 mM EDTA and 100 mM NaCl) over night at 4 °C before use. For blocking, the spread coverslips were treated with 10% NGS in TE buffer 1 (10 mM Tris–HCl pH 7.5, 1 mM EDTA) for 30 min at room temperature. After brief washing with TE buffer 1, the spread coverslips were incubated with 10% NGS, 75 nM fluorescent Py–Im polyamide probe and 0.5 μg/ml DAPI in TE buffer 1 containing 10% DMF for 1 h at 37 °C. After washing with TEN100 (10 mM Tris–HCl pH 7.5, 1 mM EDTA and 100 mM NaCl) buffer (five times for 3 min), the spread coverslips were mounted in PPDI solution (10 mM HEPES pH 7.5, 100 mM KCl, 1 mM $MgCl_2$, 80% glycerol, 1 mg/mL paraphenylene diamine) and the coverslips were sealed with a nailporish. Sectioning images were recorded with DeltaVision and projected ("Quick Projection" tool) without deconvolution to obtain telomeric signals and background signals. For quantification of the telomere signals in Fig. 4.2, the telomere signals yielded by the polyamides were extracted based on threshold values using the Softworks software (Applied Precision). The maximum intensity values of signals in the extracted telomere regions were then used as telomere signals. For the background signals, ROIs which were same shape of each telomere were set next to the defined each telomere signals in the chromosomes, and maximum values of the signals were used as background signals. Signal-to-background (S/B) ratio is calculated by dividing the signal intensities of the telomeric foci by the background intensities from its neighboring chromosomal region. Box plot and dot plot were created in R software.

Formaldehyde–fixed HeLa 1.3 Cells Treated with Fluorescent Py–Im Polyamide Probes [23]. For polyamide staining, the HeLa 1.3 cells were grown on coverslips coated with polylysine. The cell coverslips were washed in phosphate–buffered saline (PBS) twice and fixed with 1.85% formaldehyde in PBS for 15 min at room temperature. The fixed cell coverslips were then treated with 50 mM glycine in PBS for 5 min, and permeabilized with 0.5% Triton X–100 in PBS for 5 min. After briefly washing with HMK buffer (10 mM HEPES pH 7.5, 1 mM $MgCl_2$, 100 mM KCl) twice, the coverslips were soaked in HEN buffer for overnight and kept at 4 °C until use. For blocking, the cell coverslips were treated with 10% NGS in TE buffer 1 for 30 min at room temperature. After brief washing with TE buffer 1, the cell coverslips were incubated with 10% NGS, 15 nM fluorescent Py–Im polyamide probe and 0.5 μg/ml DAPI in TE buffer 1 containing 10% DMF for 1 h at 37 °C. After washing with TEN200 buffer (10 mM Tris–HCl pH 7.5, 1 mM EDTA, and 200 mM NaCl) (five times for 3 min), the mounting and subsequent image acquisitions were performed as described above. For quantification of the telomere signals in Fig. 4.3, they were extracted from the images as described previously [23]. The telomere signals yielded by the polyamides were extracted based on threshold values using the Softworks software (Applied Precision). The maximum intensity values of signals in the extracted telomere regions were then used as telomere signals. For the background signals, 10 squares (10 pixels × 10 pixels) were randomly set outside the defined telomere signals and mean values of the signals in the squares were used as background signals. Signal-to-background (S/B) ratio is calculated described above. Box plot and dot plot were created in R software.

4.4.5 Identification of TDi59-A and TTet59-B Binding Sites in Genome

Pull–down and Affinity Purification of DNA Bound by Biotin TDi59–A and Biotin TTet59–B [16d, f, g]. Human foreskin BJ fibroblast cells were maintained in 8 mL DMEM medium containing 10% FBS, 1x non–essential amino acids, penicillin (100 units/mL) and streptomycin (100 μg/mL) at 37 °C (5% CO_2). 75–80% confluent cells (2×10^6 cells) were washed with PBS and isolated by tripsinization for 3 min. 3 plates of 2×10^6 cells were used. Each isolated 2×10^6 cells were again washed with cold PBS and the cell pellet was suspended in 5 mL ice cold NP–40 lysis buffer (10 mM Tris–HCl pH 7.4, 10 mM NaCl, 3 mM $MgCl_2$, 0.5% Nonidet P–40, 0.15 mM spermine, 0.5 mM spermidine and 0.015x protease inhibitor cocktail) and incubated on ice for 5 min. The nuclei were pelleted by centrifugation at 300 g for 10 min and then carefully dissolved in 900 μL binding buffer (10 mM Tris–HCl pH 8.0, 5 mM $MgCl_2$, 1 mM Dithiothreitol, 0.3 M KCl, 0.3x protease inhibitor cocktail and 10% glycerol) [16f] and incubated at 4 °C for 1 h. 100 μL of the binding buffer containing 1 μL of 400 μM biotinylated Py–Im polyamide probe dissolved in DMSO was added to the extracted nuclei solution. Each 2×10^6 cells nuclei were incubated with biotinylated Py–Im polyamide probe (400 nM final concentration) dissolved in DMSO (0.1% final concentration) at 4 °C for 16 h. After incubation, the nuclei were pelleted by centrifugation at 5200 rpm for 10 min and then washed with 1 mL washing buffer A (10 mM Tris–HCl pH 7.4, 15 mM NaCl, 60 mM KCl, 0.15 mM spermine, 0.5 mM spermidine and 0.1x protease inhibitor cocktail). After removal of washing buffer A by centrifuge at 5200 rpm for 10 min, the pellets were resuspended in 50 μL MNase buffer (20 mM Tris–HCl pH 8.0, 5 mM NaCl, 2.5 mM $CaCl_2$). Then 4 units micrococcal nuclease (MNase) in 50 μL MNase buffer containing 43 μg RNase A and 0.2x protease inhibitor cocktail was added to the solution, followed by incubation at 37 °C for 30 min to digest to mononucleus [16f]. After this digestion 2 μL of 0.5 M EDTA was added to quench MNase and then histone protein was removed by addition of 4 μL of 20 mg/mL proteinase K solution and incubation at 37 °C for 30 min. After MNase digestion and proteinase K treatment, the suspension was mixed with equal volume of modified COSMIC [4a] buffer (20 mM Tris–HCl pH 8.1, 2 mM EDTA, 150 mM NaCl, 0.3x protease inhibitor cocktail, 1% Triton–X 100, and 0.1% SDS) containing 0.50 mg of suspended streptavidin–coated magnetic beads (C1 beads) and then incubated at 4 °C for 20 h to pull down target DNA. Preparation of the beads was performed as follows: after removing the suspension solution from Dynabeads® MyOne™ Streptavidin C1, they were washed with modified COSMIC buffer and resuspended in it.

After the pull down, beads–bound DNA was isolated by reported affinity purification [16d, f, g]. Beads were washed once with 500 μL of washing buffer 1 (10 mM Tris–HCl pH 8.0, 1 mM EDTA, 0.3% SDS) for 7 min, once with 500 μL of washing buffer 2 (10 mM Tris–HCl pH 8.0, 250 mM LiCl, 1 mM EDTA, 0.5% Nonidet P–40) for 5 min, twice with 500 μL washing buffer 3 (10 mM Tris–HCl pH 7.5, 1 mM EDTA, 0.1% Nonidet P–40) for 5 min and twice with 500 μL TE buffer 2 (10 mM Tris–HCl pH 8.0 and 1 mM EDTA) for 5 min. Beads were resuspended in elution buffer 1 (10 mM Tris–HCl pH 7.6, 0.4 mM EDTA and 100 mM KOH) and DNA was eluted from magnetic beads after heating at 90 °C for 30 min. The remaining DNA with the beads was eluted using elution buffer 2 (2% SDS, 100 mM $NaHCO_3$ and 3 mM biotin) with the heating at 65 °C for 8 h. The detached DNA was purified with QIAquick PCR Purification Kit following the manufacture protocol and then quantified with a Nanodrop. Isolated DNA were pooled to get optimum DNA concentration for the sequencing library construction. Biological replicates were assessed.

High–throughput Sequencing [16f]. Sequencing libraries of the collected DNA were prepared with standard Ion Xpress^{TM} Plus gDNA Fragment Library Preparation reagents (Thermo Fisher Scientific Inc.). A1 and P1 adapter–ligated enriched DNA was amplified and purified, followed by quality and quantity check with Agilent 2100 Bioanalyzer. The qualified libraries were used for high–throughput sequencing. Following the manufacture's protocol, the sequencing was performed, starting with template preparation using Ion PI^{TM} template OT2 200 kit in Ion one touch2 system (Thermo Fisher Scientific Inc.). The templates were then enriched using Ion one touch ES. The enriched libraries were sequenced with 300 flow of single read by Ion Proton^{TM} Sequencer using Ion PI^{TM} Sequencing 200 kit v3/Ion PI chip by following the manufacturer's guidelines. 25–30 million post filtered reads per library were produced. The data were handled by employing standard program packages in the Ion torrent suit. Torrent Mapping Alignment Program 4.4.2 (TMAP) was used for aligning reads with the hg38 reference genome, and enriched peaks was called using MACS 1.4.2 [30a]. The enriched genomic regions were visualized using UCSC genome browser (http://genome.ucsc.edu/). Homer motif analysis program was used for the de novo motif analysis [30b].

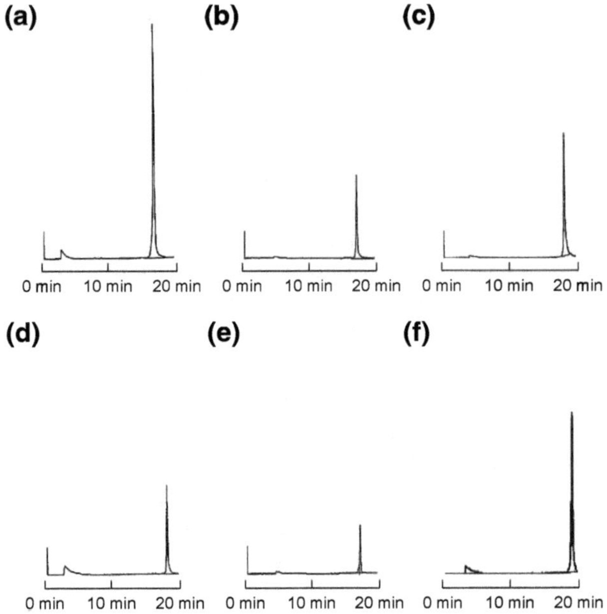

Fig. S4.1 Analytical HPLC profiles of Py–Im polyamides (**a**: **TTet59–A**, **b**: **TTet59–B**, **c**: **TAMRA TTet59–A**, **d**: **TAMRA TTet59–B**, **e**: **Biotin TDi59–A**, **f**: **Biotin TTet59–B**). Copyright 2016 American Chemical Society

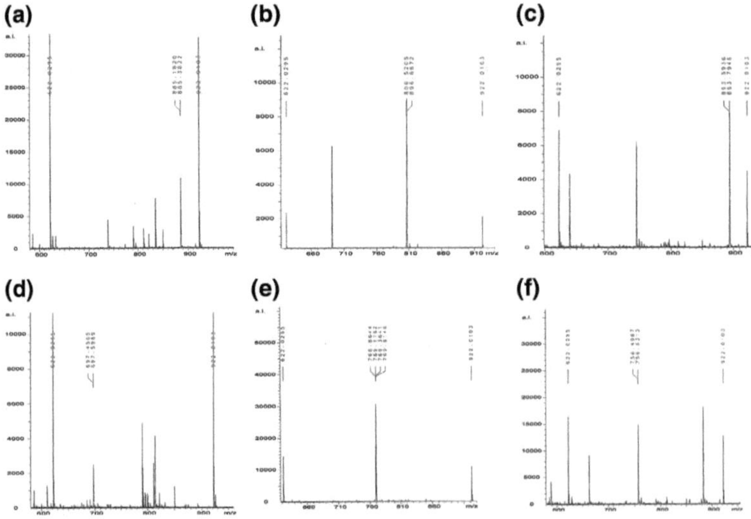

Fig. S4.2 ESI–TOF MS spectra of Py–Im polyamides (**a**: **TTet59–A**, **b**: **TTet59–B**, **c**: **TAMRA TTet59–A**, **d**: **TAMRA TTet59–B**, **e**: **Biotin TDi59–A**, **f**: **Biotin TTet59–B**). Copyright 2016 American Chemical Society

ODN1: 5'-Biotin-GGTT**AGGGTTAGGGTTAGGGTTAGGGTT**AGGTT
TT
3'-CCAA**TCCCAATCCCAATCCCAATCCCAA**TCC

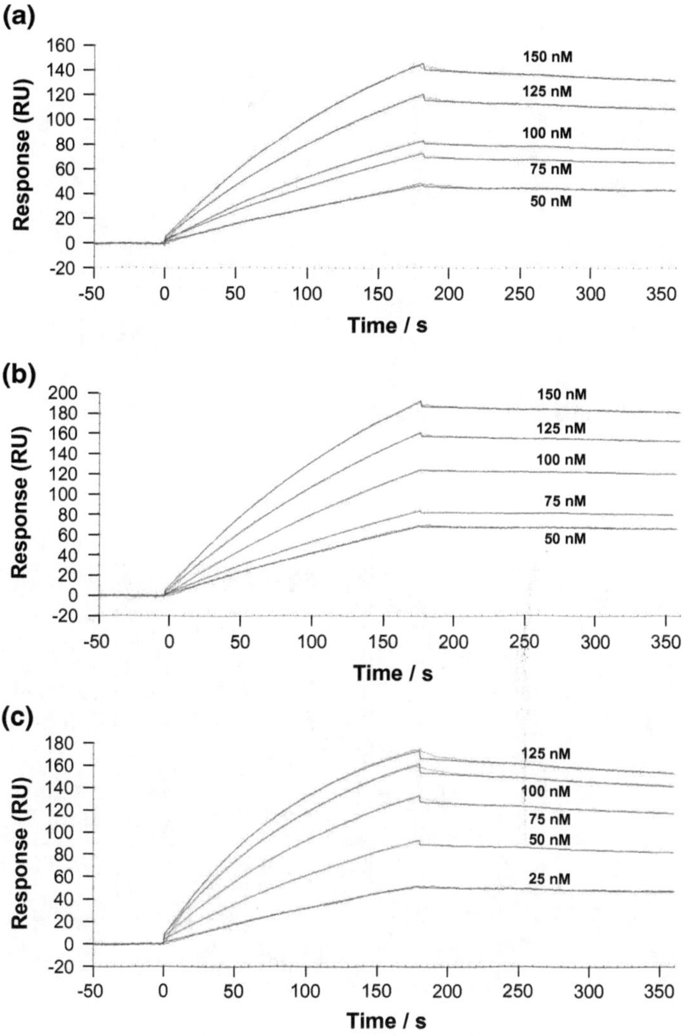

Fig. S4.3 SPR sensorgrams for the interaction between ODN-1 and (**a**) **TTet59–A**, (**b**) **TTet59–B** or (**c**) **TTri59–A**. Copyright 2016 American Chemical Society

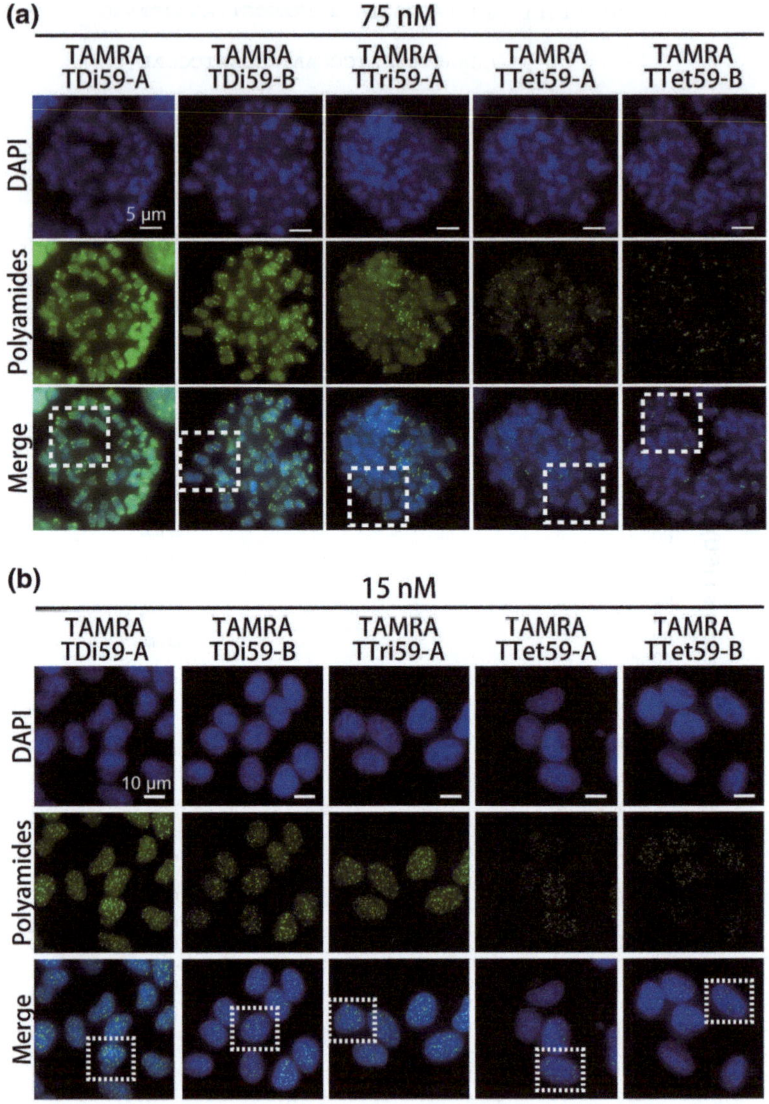

Fig. S4.4 Full images of HeLa 1.3 cell spreads in Fig. 4.2 and HeLa 1.3 cells in Fig. 4.3. Enlarged images of the boxed regions in (**a**) and (**b**) are shown in Fig. 4.2 and 3, respectively. Copyright 2016 American Chemical Society

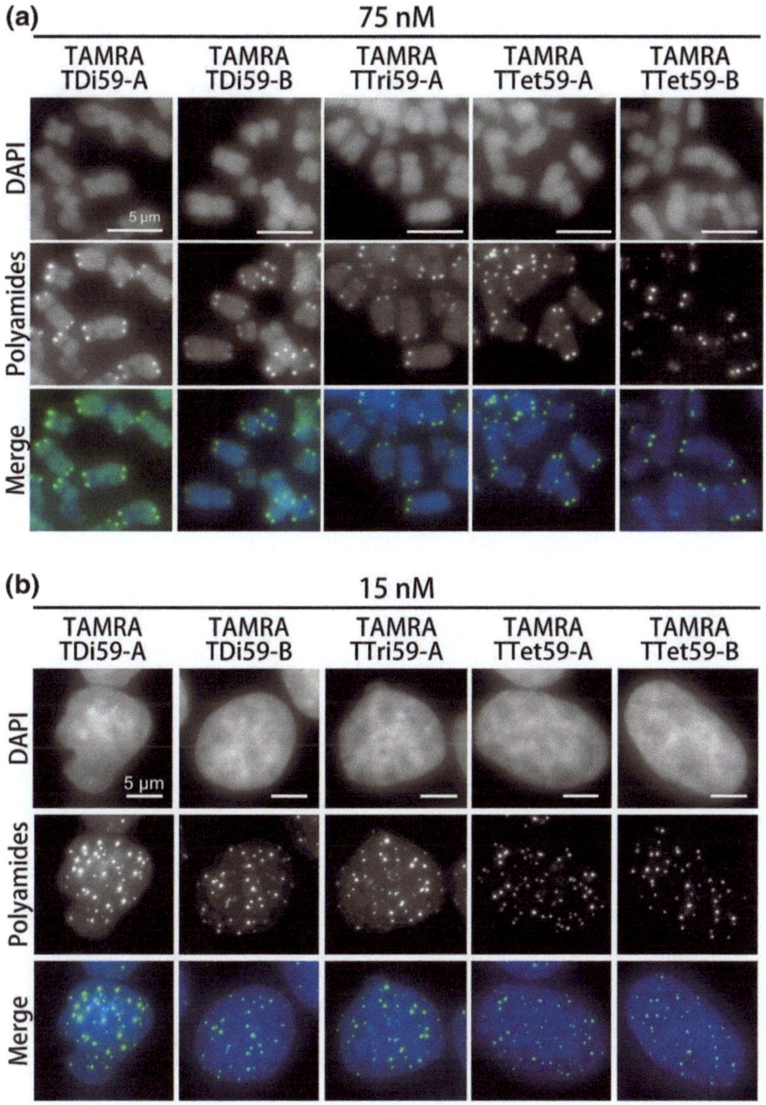

Fig. S4.5 Grey-scale and color merged (DNA, blue; polyamides, green) images of (**a**) HeLa 1.3 cell spreads in Fig. 4.2 and (**b**) HeLa 1.3 cells in Fig. 4.3. Copyright 2016 American Chemical Society

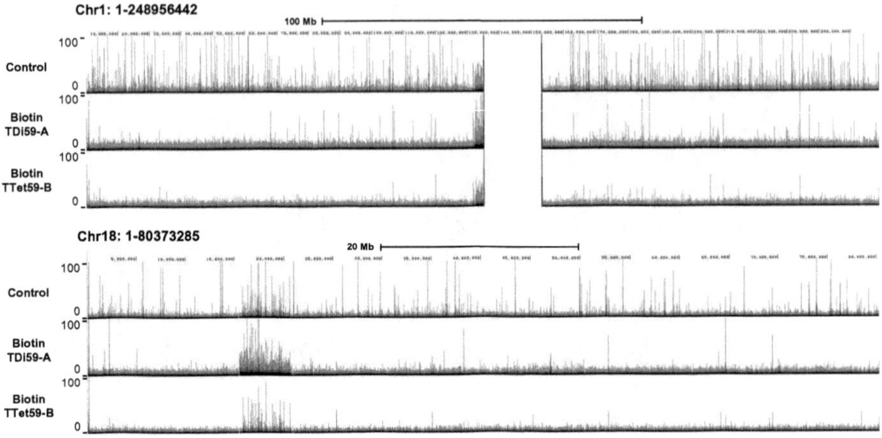

Fig. S4.6 Mapping of the biotinylated Py–Im polyamide probes binding and enriched sites in the whole of chromosomes 1 and 18. Copyright 2016 American Chemical Society

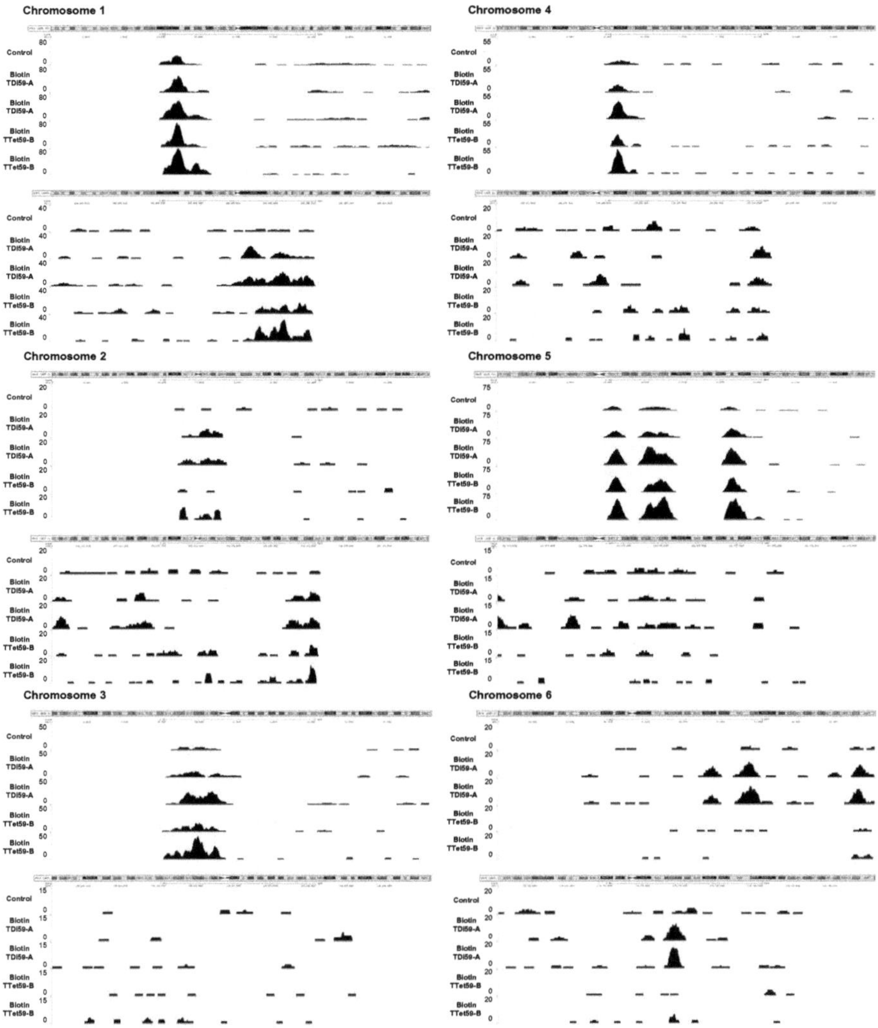

Fig. S4.7 The enrichment of **Biotin TDi59–A** and **Biotin TTet59–B** at both termini for all chromosomes Copyright 2016 American Chemical Society

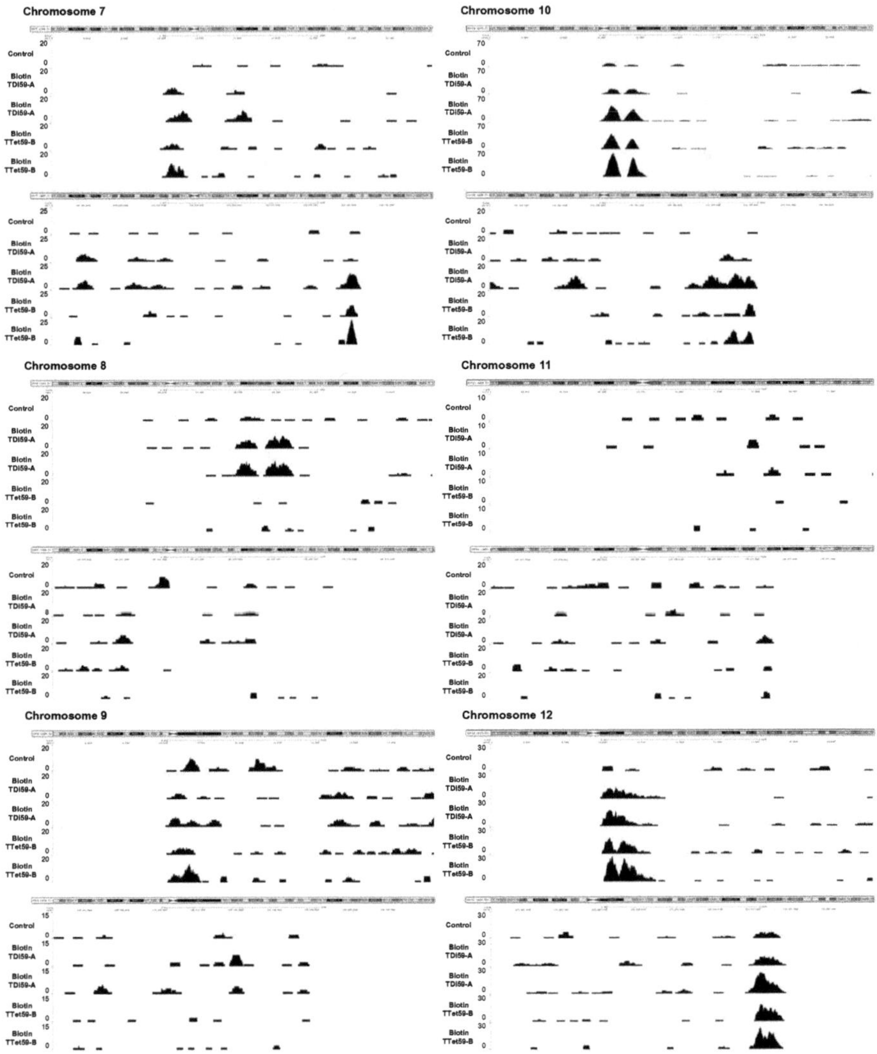

Fig. S4.7 (continued)

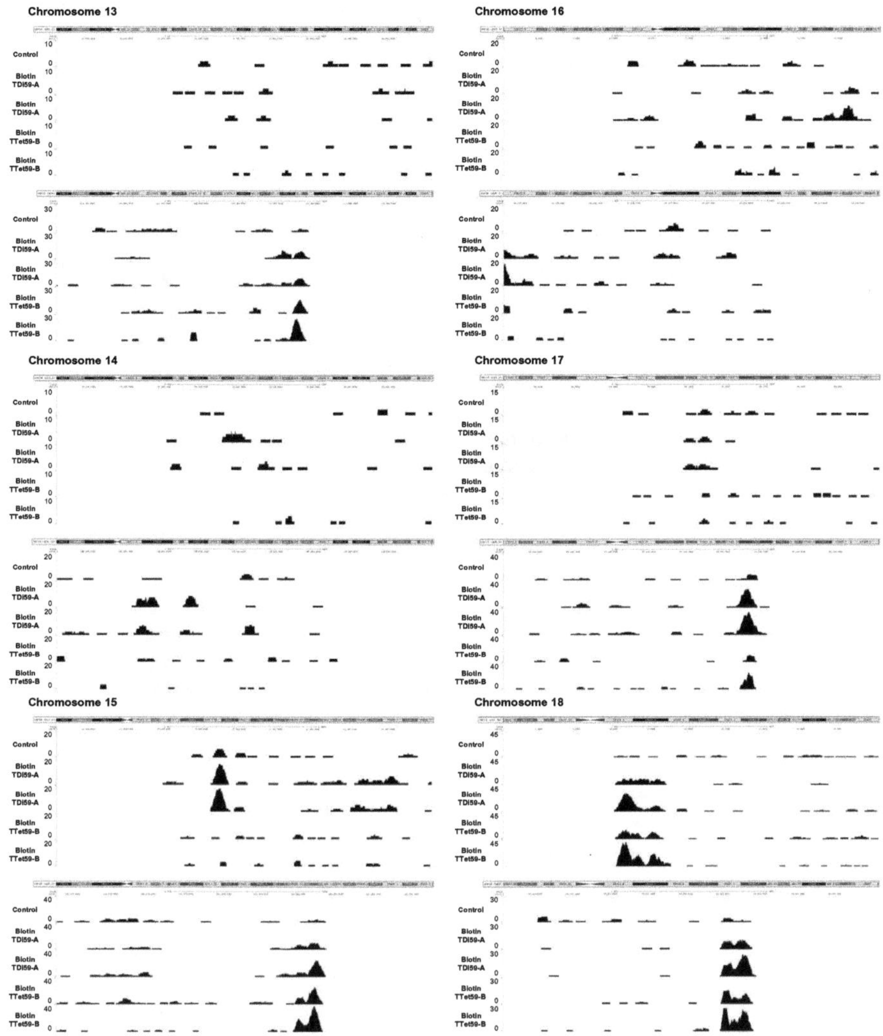

Fig. S4.7 (continued)

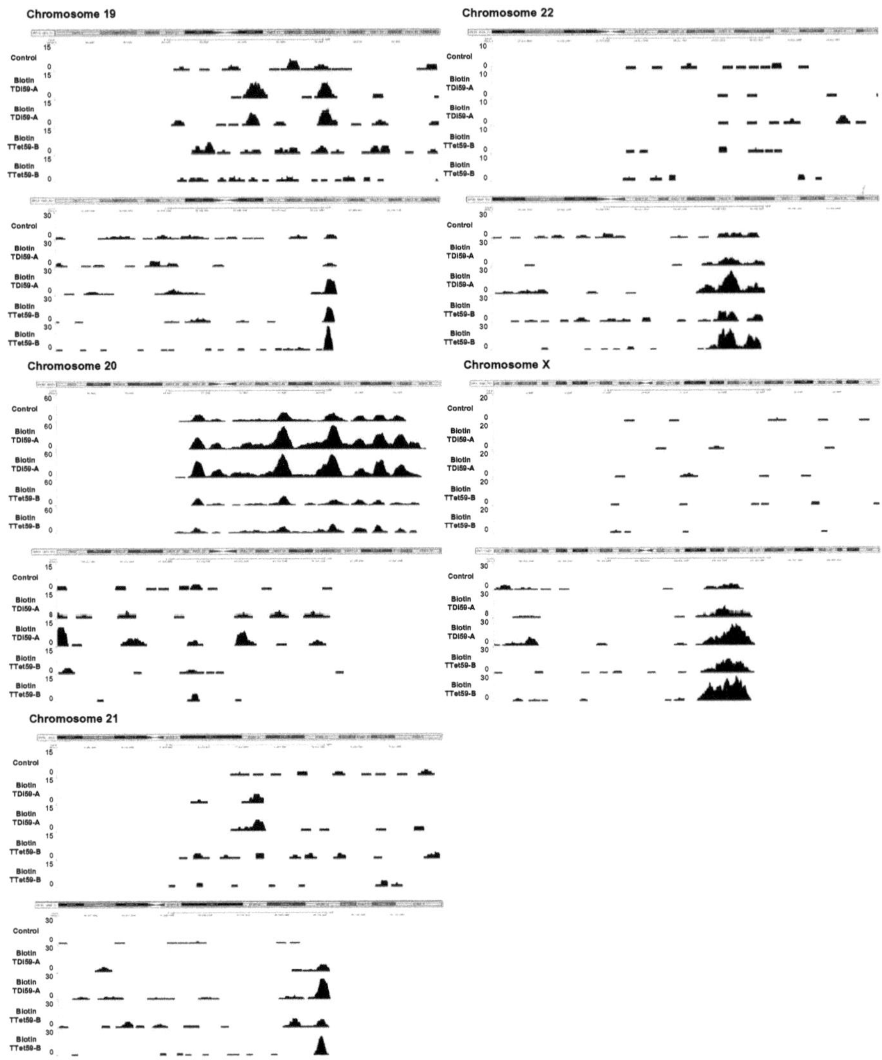

Fig. S4.7 (continued)

References

1. Jinek M, Chylinski K, Fonfara I, Hauer M, Doudna JA, Charpentier E (2012) Science 337:816–821
2. Klug A (2010) Annu Rev Biochem 79:213–231
3. Boch J, Scholze H, Schornack S, Landgraf A, Hahn S, Kay S, Lahaye T, Nickstadt A, Bonas U (2009) Science 326:1509–1512
4. (a) Duca M, Vekhoff P, Oussedik K, Halby L, Arimondo PB (2008) Nucleic Acids Res 36:5123–5138. (b) Moser HE, Dervan PB (1987) Science 238:645–650
5. (a) Nielsen PE (2010) Chem Biodivers 7:786–804. (b) Nielsen PE, Egholm M, Berg RH, Buchardt O (1991) Science 254:1497–1500
6. (a) Liu Y, Chai Y, Kumar A, Tidwell RR, Boykin DW, Wilson WD (2012) J Am Chem Soc 134:5290–5299. (b) Paul A, Nanjunda R, Kumar A, Laughlin S, Nhili R, Depauw S, Deuser SS, Chai Y, Chaudhary AS, David–Cordonnier MH, Boykin DW, Wilson WD (2015) Bioorg Med Chem Lett 25:4927–4932
7. Rodríguez J, Mosquera J, García-Fandiño R, Vázquez ME, Mascareñas JL (2016) Chem Sci 7:3298–3303
8. (a) Holman GG, Zewail–Foote M, Smith AR, Johnson KA, Iverson BL (2011) Nat Chem 3:875–881. (b) Smith AR, Iverson BL (2013) J Am Chem Soc 135:12,783–12,789
9. (a) Dervan PB (2001) Bioorg Med Chem 9:2215–2235. (b) Dervan PB, Edelson BS (2003) Curr Opin Struc Biol 13:284–299. (c) Dervan PB, Doss RM, Marques MA (2005) Curr Med Chem Anti–Cancer Agents 5:373–387. (d) Bando T, Sugiyama H (2006) Acc Chem Res 39:935–944. (e) Blackledge MS, Melander C (2013) Bioorg Med Chem 21:6101–6114. (f) Trauger JW, Baird EE, Dervan PB (1996) Nature 382:559–561. (g) White S, Szewczyk JW, Turner JM, Baird EE, Dervan PB (1998) Nature 391:468–471
10. (a) Herman DM, Baird EE, Dervan PB (1998) J Am Chem Soc 120:1382–1391. (b) Swalley SE, Baird EE, Dervan PB (1999) J Am Chem Soc 121:1113–1120
11. (a) Mrksich M, Parks ME, Dervan PB (1994) J Am Chem Soc 116:7983–7988. (b) de Clairac RPL, Geierstanger BH, Mrksich M, Dervan PB, Wemmer DE (1997) J Am Chem Soc 119:7909–7916
12. (a) Herman DM, Turner JM, Baird EE, Dervan PB (1999) J Am Chem Soc 121:1121–1129. (b) Chenoweth DM, Dervan PB (2009) Proc Natl Acad Sci U S A 106:13175–13179. (c) Chenoweth DM, Dervan PB (2010) J Am Chem Soc 132:14521–14529. (d) Morinaga H, Bando T, Takagaki T, Yamamoto H, Hashiya K, Sugiyama H (2011) J Am Chem Soc 133:18924–18930. (e) Li BC, Montgomery DC, Puckett JW, Dervan PB (2013) J Org Chem 78:124–133
13. (a) Parks ME, Baird EE, Dervan PB (1996) J Am Chem Soc 118:6147–6152. (b) Turner JM, Swalley SE, Baird EE, Dervan PB (1998) J Am Chem Soc 120:6219–6226
14. (a) Baird EE, Dervan PB (1996) J Am Chem Soc 118:6141–6146. (b) Wurtz NR, Turner JM, Baird EE, Dervan PB (2001) Org Lett 3:1201–1203. (c) Fan L, Yao G, Pan Z, Wu C, Wang HS, Burley GA, Su W (2015) Org Lett 17:158–161
15. (a) Vaijayanthi T, Bando T, Pandian GN, Sugiyama H (2012) ChemBioChem 13:2170–2185. (b) Boutorine AS, Novopashina DS, Krasheninina OA, Nozeret K, Venyaminova AG (2013) Molecules 18:15357–15397. (c) Nozeret K, Loll F, Escudé C, Boutorine AS (2015) ChemBioChem 16:549–554
16. (a) Anandhakumar C, Kizaki S, Bando T, Pandian GN, Sugiyama H (2015) ChemBioChem 16:20–38. (b) Meier JL, Yu AS, Korf I, Segal DJ, Dervan PB (2012) J Am Chem Soc 134:17814–17822. (c) Kang JS, Meier JL, Dervan PB (2014) J Am Chem Soc 136:3687–3694. (d) Erwin GS, Bhimsaria D, Eguchi A, Ansari AZ (2014) Angew Chem Int Ed 53:10124–10128. (e) Anandhakumar C, Li Y, Kizaki S, Pandian GN, Hashiya K, Bando T, Sugiyama H (2014) ChemBioChem 15:2647–2651. (f) Chandran A, Syed J, Taylor RD, Kashiwazaki G, Sato S, Hashiya K, Bando T, Sugiyama H (2016) Nucleic Acids Res 44:4014–4024. (g) Chandran A, Syed J, Li Y, Sato S, Bando T, Sugiyama H (2016) ChemBioChem 17:1905–1910

17. (a) Murty MSRC, Sugiyama H (2004) Biol Pharm Bull 27:468–474. (b) Gottesfeld JM, Neely L, Trauger JW, Baird EE, Dervan PB (1997) Nature 387:202–205. (c) Lai Y-M, Fukuda N, Ueno T, Kishioka H, Matsuda H, Saito S, Matsumoto K, Ayame H, Bando T, Sugiyama H, Mugishima H, Serie K (2005) J Pharmacol Exp Ther 315:571–575. (d) Hiraoka K, Inoue T, Taylor RD, Watanabe T, Koshikawa N, Yoda H, Shinohara K, Takatori A, Sugimoto K, Maru Y, Denda T, Fujiwara K, Balmain A, Ozaki T, Bando T, Sugiyama H, Nagase H (2015) Nat Commun 6:6706
18. (a) Xiao X, Yu P, Lim HS, Sikder D, Kodadek T (2007) Angew Chem Int Ed 46:2865–2868. (b) Patel S, Jung D, Yin P, Carlton P, Yamamoto M, Bando T, Sugiyama H, Lee KB (2014) ACS Nano 8:8959–8967
19. (a) Han L, Pandian GN, Junetha S, Sato S, Chandran A, Taniguchi J, Saha A, Bando T, Nagase H, Sugiyama H (2013) Angew Chem Int Ed 52:13410–13413. (b) Pandian GN, Taniguchi J, Junetha S, Sato S, Han L, Saha A, Anandhkumar C, Bando T, Nagase H, Vaijayanthi T, Taylor RD, Sugiyama H (2014) Sci Rep 4:3843. (c) Han L, Pandian GN, Chandran A, Sato S, Taniguchi J, Kashiwazaki G, Sawatani Y, Hashiya K, Bando T, Xu Y, Qian X, Sugiyama H (2015) Angew Chem Int Ed 54:8700–8703
20. Singh I, Wendeln C, Clark AW, Cooper JM, Ravoo BJ, Burley GA (2013) J Am Chem Soc 135:3449–3457
21. (a) Trauger JW, Baird EE, Dervan PB (2014) J Am Chem Soc 120:3534–3535. (b) Yamamoto M, Bando T, Morinaga N, Kawamoto Y, Hashiya K, Sugiyama H (2014) Chem Eur J 20:752–759
22. (a) Herman DM, Baird EE, Dervan PB (1999) Chem Eur J 5:975–983. (b) Kers I, Dervan PB (2002) Bioorg Med Chem 10:3339–3349. (c) Schaal TD, Mallet WG, McMinn DL, Nguyen NV, Sopko MM, John S, Parekh BS (2018) Nucleic Acids Res 31:1282–1291. (d) Sasaki S, Bando T, Minoshima M, Shinohara K, Sugiyama H (2008) Chem Eur J 14:864–870. (e) Yamamoto M, Bando T, Kawamoto Y, Taylor R, Hashiya K, Sugiyama H (2014) Bioconjugate Chem 25:552–559. (f) Taylor RD, Kawamoto Y, Hashiya K, Bando T, Sugiyama H (2014) Chem. Asian J 9:2527–2533
23. (a) Maeshima K, Janssen S, Laemmli UK (2001) EMBO J 20:3218–3228. (b) Kawamoto Y, Bando T, Kamada F, Li Y, Hashiya K, Maeshima K, Sugiyama H (2013) J Am Chem Soc 135:16468–16477. (c) Hirata A, Nokihara K, Kawamoto Y, Bando T, Sasaki A, Ide S, Maeshima K, Kasama T, Sugiyama H (2014) J Am Chem Soc 136:11546–11554. (d) Kawamoto Y, Sasaki A, Hashiya K, Ide S, Bando T, Maeshima K, Sugiyama H (2015) Chem Sci 6:2307–2312. (e) Sasaki A, Ide S, Kawamoto Y, Bando T, Murata Y, Shimura M, Yamada K, Hirata A, Nokihara K, Hirata T, Sugiyama H, Maeshima K (2016) Sci Rep 6:29261
24. (a) Smogorzewska A, de Lange T (2004) Annu Rev Biochem 73:177–208. (b) Palm W, de Lange T (2008) Annu Rev Genet 42:301–334. (c) Blackburn EH (2010) Angew Chem Int Ed 49:7405–7421. (d) Xu Y (2011) Chem Soc Rev 40:2719–2740. (e) Nandakumar J, Cech TR (2013) Nat Rev Mol Cell Biol 14:69–82. (f) Zakian VA (2012) Exp Cell Res 318:1456–1460. (g) Smogorzewska A, van Steensel B, Bianchi A, Oelmann S, Schaefer MR, Schnapp G, de Lange T (2000) Mol Cell Biol 20:1659–1668
25. (a) Biffi G, Tannahill D, McCafferty J, Balasubramanian S (2013) Nat Chem 5:182–186. (b) Doksani Y, Wu JY, de Lange T, Zhuang X (2013) Cell 155:345–356
26. Lansdorp PM, Verwoerd NP, van de Rijke FM, Dragowska V, Little MT, Dirks RW, Raap AK, Tanke HJ (1996) Hum Mol Genet 5:685–691
27. Anders L, Guenther MG, Qi J, Fan ZP, Marineau JJ, Rahl PB, Lovén J, Sigova AA, Smith WB, Lee TI, Bradner JE, Young RA (2014) Nat Biotechnol 32:92–96
28. (a) Minoshima M, Bando T, Sasaki S, Fujimoto J, Sugiyama H (2008) Nucleic Acids Res 36:2889–2894. (b) Wetzler M, Wemmer DE (2010) Org Lett 12:3488–3490
29. (a) Lacy ER, Le NM, Price CA, Lee M, Wilson WD (2002) J Am Chem Soc 124:2153–2163. (b) Wang S, Aston K, Koeller KJ, Harris GD, Rath NP, Bashkin JK, Wilson WD (2014) Org Biomol Chem 12:7523–7536

30. (a) Feng J, Liu T, Qin B, Zhang Y, Liu XS (2012) Nat. Protoc 7:1728–1740. (b) Heinz S, Benner C, Spann N, Bertolino E, Lin YC, Laslo P, Cheng JX, Murre C, Singh H, Glass CK (2010) Mol Cell 38:576–589
31. (a) Maeshima K, Laemmli UK (2003) Dev Cell 4:467–480. (b) Maeshima K, Yahata K, Sasaki Y, Nakatomi R, Tachibana T, Hashikawa T, Imamoto F, Imamoto N (2006) J Cell Sci 119:4442–4451

Curriculum Vitae

Dr. Yusuke Kawamoto

Department of Chemistry, Graduate School of Science, Kyoto University, Kitashirakawa-Oiwakecho, Sakyo, Kyoto 606-8502, Japan

Current address

Department of Chemistry and Biochemistry, University of California, San Diego, La Jolla, California 92093-0358, United States

Email: ykawamoto@ucsd.edu

Education

- **Ph.D. in Chemistry**: April 2015 to March 2018, Department of Chemistry, Graduate School of Science, Kyoto University, Japan (Supervisor: Prof. Hiroshi Sugiyama)
- **Master of Science in Chemistry**: April 2013 to March 2015, Department of Chemistry, Graduate School of Science, Kyoto University, Japan (Supervisor: Prof. Hiroshi Sugiyama)
- **Bachelor of Science**: April 2009 to March 2013, Faculty of Science, Kyoto University, Japan

Honors and Awards

- Japan Society for the Promotion of Science Overseas Research Fellowship (July 2018 to June 2020)
- Japan Society for the Promotion of Science Overseas Challenge Program for Young Researchers (March 2018 to June 2018)

Y. Kawamoto, *Synthesis and Biological Evaluation of Pyrrole–Imidazole Polyamide Probes for Visualization of Telomeres*, Springer Theses,
https://doi.org/10.1007/978-981-13-6912-4

- Japan Society for the Promotion of Science Research Fellowship for Young Scientists (DC1) (April 2015 to March 2018)
- Kyoto University Short Term Visit Grant for International Collaborative Research (October 2016 to November 2016)
- WPI-iCeMS Overseas Visit Grant for Young Researchers (June 2015)
- Holy Poster Prize at the 23rd International Roundtable on Nucleoside, Nucleotides and Nucleic Acids (August 29, 2018)
- The Chemical Society of Japan (CSJ) Student Presentation Award 2017 at the 97th Annual CSJ Meeting (March 30, 2017)

Research Interest

I am interested in chemical biology, in which especially nucleic acids attract me. My research aims to develop fluorescent probes to label nucleic acids. As a doctoral candidate I worked on DNA binding molecules to visualize specific regions. Now I am working on fluorescent nucleic acids to study secondary structures and modification reactions of DNA and RNA.

MIX
Papier aus verantwortungsvollen Quellen
Paper from responsible sources
FSC® C105338

If you have any concerns about our products,
you can contact us on
ProductSafety@springernature.com

In case Publisher is established outside the EU,
the EU authorized representative is:
Springer Nature Customer Service Center GmbH
Europaplatz 3, 69115 Heidelberg, Germany

Printed by Libri Plureos GmbH
in Hamburg, Germany